# Orbital Angular Momentum States of Light

## Propagation through atmospheric turbulence

# IOP Series in Advances in Optics, Photonics and Optoelectronics

## SERIES EDITOR

**Professor Rajpal S Sirohi** Consultant Scientist

## About the Editor

Rajpal S Sirohi is currently working as a faculty member in the Department of Physics, Alabama A&M University, Huntsville, Alabama (USA). Prior to this, he was a consultant scientist at the Indian Institute of Science Bangalore, and before that he was chair professor in the Department of Physics, Tezpur University, Assam. During 2000–11, he was academic administrator, being vice chancellor to a couple of universities and the director of the Indian Institute of Technology Delhi. He is the recipient of many international and national awards and the author of more than 400 papers. Dr Sirohi is involved with research concerning optical metrology, optical instrumentation, holography, and speckle phenomenon.

## About the series

Optics, photonics and optoelectronics are enabling technologies in many branches of science, engineering, medicine and agriculture. These technologies have reshaped our outlook, our way of interaction with each other and brought people closer. They help us to understand many phenomena better and provide a deeper insight in the functioning of nature. Further, these technologies themselves are evolving at a rapid rate. Their applications encompass very large spatial scales from nanometers to astronomical and a very large temporal range from picoseconds to billions of years. The series on the advances on optics, photonics and optoelectronics aims at covering topics that are of interest to both academia and industry. Some of the topics that the books in the series will cover include bio-photonics and medical imaging, devices, electromagnetics, fiber optics, information storage, instrumentation, light sources, CCD and CMOS imagers, metamaterials, optical metrology, optical networks, photovoltaics, freeform optics and its evaluation, singular optics, cryptography and sensors.

## About IOP ebooks

The authors are encouraged to take advantage of the features made possible by electronic publication to enhance the reader experience through the use of colour, animation and video, and incorporating supplementary files in their work.

## Do you have an idea of a book you'd like to explore?

For further information and details of submitting book proposals see iopscience.org/books or contact Ashley Gasque on Ashley.gasque@iop.org.

# Orbital Angular Momentum States of Light

Propagation through atmospheric turbulence

**Kedar Khare, Priyanka Lochab and Paramasivam Senthilkumaran**
*Department of Physics, Indian Institute of Technology Delhi, Hauz Khas,
New Delhi, India*

**IOP** Publishing, Bristol, UK

ISBN    978-0-7503-2280-5 (ebook)
ISBN    978-0-7503-2278-2 (print)
ISBN    978-0-7503-2281-2 (myPrint)
ISBN    978-0-7503-2279-9 (mobi)

DOI    10.1088/978-0-7503-2280-5

Version: 20201201

IOP ebooks

British Library Cataloguing-in-Publication Data: A catalogue record for this book is available from the British Library.

Published by IOP Publishing, wholly owned by The Institute of Physics, London

IOP Publishing, Temple Circus, Temple Way, Bristol, BS1 6HG, UK

US Office: IOP Publishing, Inc., 190 North Independence Mall West, Suite 601, Philadelphia, PA 19106, USA

# Contents

# Preface

This book has origins in the present authors' work over a number of years on a range of topical areas including complex signal representation in optics, propagation characteristics of phase and polarization singularities, understanding the near core structure of optical vortices, quantitative phase imaging and light propagation in random media. Over the last five years, one of the authors (PL) worked under the supervision of the other two (KK and PS) for her PhD thesis at Indian Institute of Technology Delhi, India and our lively interactions allowed us to have a re-look at these interesting topics through a beginning researcher's viewpoint. Laser beam propagation in turbulence is a mature area with literature spanning more than 70 years. A large number of publications and some good books have been devoted to this topic that is of immense interest to a number of practical applications like defense systems and free space classical and quantum communication.

PL's recently completed thesis work explored the question of whether one can utilize the degrees of freedom such as amplitude, phase, polarization, and orbital angular momentum (OAM) of light for designing laser beams that can inherently maintain robust intensity profiles on long range propagation through turbulence. While investigating this question we gained several valuable insights. At the same time we constantly felt the need for a single book/monograph that explained the key details regarding modeling of optical beam propagation through atmospheric turbulence and at the same time delved into more exotic topics like OAM states of light and polarization singularities that are essential for understanding generation and applications of structured light beams. So when it was time for PL to put together her PhD thesis document, we took up the exercise of writing this book in parallel. We believe that the material in this book will be helpful to optics students, researchers and engineers working on long range laser propagation systems in two important ways. It will give them a historical perspective on our current understanding of light propagation through turbulence, and at the same time provide computer modeling tools for making realistic assessment of arbitrary beam profiles after long range propagation. The pieces of computer code explained in the appendix of this book can be run in an open source environment like GNU Octave and may also be readily modified to suit individual systems and turbulence conditions. So in some sense this book is also an attempt to de-mystify the topic of beam propagation in turbulence and allow beginning researchers to get a quick start into the real problems they are interested to address. We welcome readers to try out the codes and contribute towards improving them further. KK would like to thank IIT Delhi administration for allowing a sabbatical year that allowed him to focus on writing this book without too many administrative distractions. PL would like to acknowledge the fellowship support and the open research environment at IIT Delhi for pursuing her recently concluded PhD thesis work. We deeply appreciate the encouragement from Professor Rajpal S Sirohi, the editor of this IOP series, for writing this book. We are also grateful for the professional support and patience shown by IoP editorial office, particularly Ms Ashley Gasque, while the manuscript of this book was being prepared. Two PhD students at our institute—Ms Ruchi and Ms Gauri Arora—gave a thorough reading to the proof copy of this book and we would like to thank them for pointing out several errors. The book was entirely written using the collaborative LaTeX environment provided by Overleaf to which our institute subscribes. Finally we are deeply indebted to our respective families for their continuous support and patience during this book writing effort.

Kedar Khare
Priyanka Lochab
Paramasivam Senthilkumaran
New Delhi
September 2020

# Author biographies

## Kedar Khare

Kedar Khare is currently an Associate Professor in the Department of Physics, Indian Institute of Technology Delhi, India. Prior to joining the faculty of IIT Delhi in December 2011, he spent several years working as a Lead Scientist in the Imaging Technologies division at General Electric Global Research, Niskayuna, USA. Professor Khare's research interests span wide ranging problems in computational optical imaging. In recent years he has been actively engaged in pursuing research work on quantitative phase imaging, Fourier phase retrieval, computational microscopy, diagnostic imaging, cryo-EM imaging and structured light propagation in turbulence. Professor Khare also serves as an Associate Editor for *Journal of Modern Optics* published by Taylor & Francis, UK. Professor Khare received his Integrated MSc in Physics from Indian Institute of Technology Kharagpur with highest honors and his PhD in Optics from The Institute of Optics, University of Rochester, USA.

## Priyanka Lochab

Priyanka Lochab is currently a Research Associate in the Department of Physics, Indian Institute of Technology Delhi, India. Her research career in optics began in 2013 when she joined IIT Delhi as a graduate student. Her major research interests are singular optics, beam propagation through random media and computational imaging.

## Paramasivam Senthilkumaran

Dr P Senthilkumaran is currently working as a full Professor in the Department of Physics, Indian Institute of Technology Delhi (IIT Delhi). He joined IIT Delhi as an Assistant Professor of Physics in 2002 and became Associate Professor in 2008 and Professor in 2012. He worked as Assistant Professor between 1996 and 2001 and as a lecturer between 1995 and 1996 in IIT Guwahati (IITG). Prior to joining IITG, he had been a senior project officer from September 1993 at IIT Madras from where he received his PhD in 1995. He is a recipient of the Young Scientist Award from Indian National Science Academy (INSA), New Delhi and Alexander von Humboldt fellowship, Germany in 1997 and 2001 respectively. He was at the University of Strathclyde, Glasgow, United Kingdom on a Royal Society, London and INSA exchange fellowship in 1999 and in Friedrich Schiller University of Jena, Germany during 2001–2002 on a Humboldt fellowship.

He has been teaching undergraduate and postgraduate courses on basic physics, electromagnetic theory, optics and lasers, Fourier optics, holography and its applications, optical metrology and optical instrumentation. His research interests are pptical beam shaping, pptical phase singularities, Berry and Pancharatnam topological phases, fiber optics, holography, non-destructive testing techniques, shear interferometry, Talbot interferometry, speckle metrology and non-linear optics. He has authored/coauthored more than 100 research publications. He has authored one book entitled *Singularities in Physics and Engineering* published by IOP Publishing, Bristol, UK in 2018. He was a guest editor of a special issue on singular optics in the *International Journal of Optics* in 2012 along with Professor Shunichi Sato and Professor Jan Masajada. Professor Senthilkumaran has also been an editorial board member of *International Journal of Optics* since 2013 and associate editor of *Optical Engineering* since 2020.

**IOP** Publishing

# Orbital Angular Momentum States of Light
### Propagation through atmospheric turbulence

**Kedar Khare, Priyanka Lochab and Paramasivam Senthilkumaran**

# Chapter 1

# Introduction

The orbital angular momentum (OAM) states of light have gained prominence over the last few decades as a promising choice of modes for free space communication as well as in defense systems. These applications inherently require long range propagation of light beams through fluctuating atmosphere. Various local factors such as temperature, pressure, humidity, wind velocity, etc, lead to space and time varying refractive index profiles that are nearly impossible to control over distances of kilometers. Even small changes in local refractive index ($\approx 10^{-5}$–$10^{-6}$) can have severe undesirable effects on a carefully designed laser beam after long range propagation through the atmosphere. Understanding and modeling of propagation of light beams with arbitrary amplitude, phase and polarization profiles through a given level of atmospheric turbulence is, therefore, of utmost importance for practical system designers.

The literature on laser beam propagation through atmospheric turbulence is spread over more than 70 years. In recent decades, several exciting developments have happened in this area in the context of advanced active beam correction methods employing adaptive optics, potential use of structured light beams for robust beam engineering, the possibility of using entangled states of light for long range communication with satellites, to name a few. The use of OAM states has been an important common feature in these developments thus creating renewed interest in the detailed study of light propagation through atmospheric turbulence. For a beginning researcher at graduate student level as well as for a seasoned optical engineer, sifting through the vast literature in this topical area is a daunting task. Most prior results in this area have been reported for Gaussian beams, although the literature on structured light propagation in atmosphere is growing. Further, testing of newer beam engineering ideas in this topical area is not easy as there is a lack of standalone, easily accessible and inexpensive software tools in the public domain that allow rigorous modeling of the propagation of beams with arbitrary field profiles through turbulence. The aim of this book and the accompanying software code pieces is to address these difficulties and make the subject accessible to a larger scientific community.

The book is organized in two main parts. In the first part we provide discussion on the basic mathematical tools used throughout the book. Chapter 2 discusses topics such as Fourier and Hilbert transforms, the basics of random processes and inverse probability law for generating realizations of random processes with given spectral properties. An important result which generalizes the notion of Hilbert transform to two-dimensions is also presented in chapter 2 which forms the basis of the novel robust beam engineering principle described later. One important aspect of this book is that for the purpose of field propagation calculations we extensively use the angular spectrum approach which we believe is better than the more commonly used Fresnel propagation method. The angular spectrum method is described in detail in chapter 3. As an illustration of the angular spectrum method, we discuss the propagation characteristics of a phase vortex with particular attention to its near-core structure in chapter 4. Propagation characteristics of the OAM states as well as vector beams representing polarization singularities through free space are described in chapters 5 and 6 respectively. Polarization singularities are often presented in optics literature as exotic objects. However, we present this topic in a more accessible manner and develop it simply as superposition of OAM states in orthogonal polarizations. This point of view is useful when discussing diversity mechanisms associated with the propagation of structured light in atmospheric turbulence.

The second part of the book deals with the description of optical properties of fluctuating atmosphere. The historical developments on modeling of wave propagation in turbulence are described in chapter 7 with reference to the Born and Rytov approximations and the extended Huygens–Fresnel principle. The numerical method for simulating beam propagation through atmosphere by means of a multiple phase screen model is described in chapter 8. In these two chapters, we provide important details on various spectrum profiles (Kolmogorov, von Karman, etc), sampling criteria in terms of atmospheric fluctuation scales, consideration on choosing number random phase of screens depending on the turbulence strength, and inclusion of sub-harmonic components to phase screens. The discussion is presented in a way that readers can readily implement these aspects in their work. Finally in chapter 9 we describe simulations and experimental results on propagation of optical beams through long range turbulence. This material is primarily based on recent work of the authors. Both scalar and vector beams are considered, with particular attention paid to the propagation of polarization singularities through turbulence. The work of the authors on engineering of polarization singular beams that can maintain a robust intensity profile on propagation through turbulence is discussed in this context by using speckle diversity as a guiding principle. Future directions and experiments that can effectively utilize the material presented in this book are also discussed. We describe the rationale behind traditionally used beam quality parameters such as scintillation indexes and their limitations in studying the characteristics of structured beams and further introduce instantaneous beam SNR as a new beam quality measure that may be more suited to understanding propagation of structured light beams through turbulence. Several pieces of annotated computer code with brief explanations are provided in appendix A. The reader can quickly incorporate or modify the computer codes as suited to their specific system modeling needs depending on the turbulence conditions of interest.

The material is not presented in terms of the coherent mode representation of optical fields as has been done in some other literature in this topic mainly because the corresponding interpretations are sometimes harder for beginning researchers or practicing optical engineers. Also, not much space is devoted to communication using OAM modes as this is not the main thrust of this book. However, we believe that the researchers involved in work on free space (classical as well as quantum) optical communication will find the details of turbulence propagation modeling useful for evaluating their new ideas. The material covered in this book along with an extensive survey of the literature and computer codes can be covered as a special topics graduate level course or can simply be treated as reference material. The simulation codes are consciously written in a way that they can be used with a proprietary software like MATLAB as well as with its open source clone GNU Octave. The codes do not use any software specific features and can be quickly adapted to model a number of practical long range propagation scenarios without difficulty. We believe that with these features, the book will be useful to a number of researchers and optical system engineers.

# Orbital Angular Momentum States of Light
Propagation through atmospheric turbulence
**Kedar Khare, Priyanka Lochab and Paramasivam Senthilkumaran**

# Chapter 2

## Mathematical preliminaries

In this chapter we describe some basic mathematical tools that will be used throughout the book. The material covered here will also help to set the basic notations. The topics discussed here include results on the Fourier transform and random processes that are traditionally covered in detail in excellent textbooks. Some specialized results such as quadrature representation for two-dimensional signals will also be discussed in detail as they will be useful in later chapters of the book.

## 2.1 Fourier transform basics

Fourier transform is an important tool that we will be using often in this book in the context of propagation of wave fields. For a finite energy function $g(x, y)$, we will use the following definition for the Fourier transform pair:

$$G(f_x, f_y) = \mathcal{F}[g(x, y)] = \int_{-\infty}^{\infty} \int_{-\infty}^{\infty} dx \, dy \, g(x, y) \exp[-i2\pi(f_x x + f_y y)], \quad (2.1)$$

and

$$g(x, y) = \mathcal{F}^{-1}[G(f_x, f_y)]$$
$$= \int_{-\infty}^{\infty} \int_{-\infty}^{\infty} df_x \, df_y \, G(f_x, f_y) \exp[i2\pi(f_x x + f_y y)]. \quad (2.2)$$

The definition in equations (2.1) and (2.2) suggests that the function $g(x, y)$ is expressible as a linear combination of complex exponentials, with the Fourier transform $G(f_x, f_y)$ denoting the weight corresponding to the term $\exp[i2\pi(f_x x + f_y y)]$. Some of the well-known properties of the Fourier transform that we will find useful are:

1. **Linearity**: the Fourier transform is a linear operator, thus, for constants $c_1$ and $c_2$,

$$\mathcal{F}[c_1 \, g_1(x, y) + c_2 \, g_2(x, y)] = c_1 \, G_1(f_x, f_y) + c_2 \, G_2(f_x, f_y). \tag{2.3}$$

2. **Scaling**: the Fourier transform of the scaled function $g(ax, by)$ is given by:

$$\mathcal{F}[g(ax, by)] = \frac{1}{|a||b|} G\left(\frac{f_x}{a}, \frac{f_y}{b}\right). \tag{2.4}$$

The stretching of a function in $(x, y)$ space thus causes compression of its Fourier transform and vice versa.

3. **Shift**: the Fourier transform of a shifted function $g(x - a, y - b)$ is given by:

$$\mathcal{F}[g(x - a, y - b)] = G(f_x, f_y)\exp[-i2\pi(f_x a + f_y b)]. \tag{2.5}$$

Translation of a function is thus equivalent to multiplication of its Fourier transform by a phase ramp.

4. **Energy theorem**: the energy of a function in $(x, y)$ space is conserved in the two-dimensional Fourier space. Thus,

$$\int_{-\infty}^{\infty} \int_{-\infty}^{\infty} dx \, dy \, |g(x, y)|^2 = \int_{-\infty}^{\infty} \int_{-\infty}^{\infty} df_x \, df_y \, |G(f_x, f_y)|^2. \tag{2.6}$$

5. **Convolution**: the convolution operation of two functions in $(x, y)$ space is equivalent to the product of their Fourier transforms. We will often use this property later in the book when describing free space diffraction of a beam.

$$\mathcal{F}\left[\int_{-\infty}^{\infty} \int_{-\infty}^{\infty} dx' \, dy' \, g(x', y')h(x - x', y - y')\right] = G(f_x, f_y)H(f_x, f_y). \tag{2.7}$$

These properties have been stated here without proofs. The proofs can be found elsewhere in excellent resources on Fourier transform theory [1, 2]. For completeness we will introduce the notation for delta impulse which may be represented as:

$$\delta(x, y) = \int_{-\infty}^{\infty} \int_{-\infty}^{\infty} df_x \, df_y \, \exp[i2\pi(f_x x + f_y y)]. \tag{2.8}$$

The delta impulse is a generalized function or distribution and the sampling property associated with it is given by:

$$\int_{-\infty}^{\infty} \int_{-\infty}^{\infty} dxdy \, \delta(x - x_0, y - y_0)g(x, y) = g(x_0, y_0). \tag{2.9}$$

Multiplication of $g(x, y)$ with a delta impulse located at $x = x_0$, $y = y_0$ followed by integration over the whole $(x, y)$ plane thus provides the value of the sample $g(x_0, y_0)$ of the function $g(x, y)$.

A signal $g(x, y)$ is called band-limited if its Fourier transform $G(f_x, f_y)$ is zero outside some spatial frequency interval $2\Omega_x \times 2\Omega_y$. Practically, most signals will have negligible energy outside a certain frequency band. Band-limited signals admit a sampling expansion of the form:

$$g(x, y) = \sum_{m=-\infty}^{\infty} \sum_{n=-\infty}^{\infty} g\left(\frac{m}{2\Omega_x}, \frac{n}{2\Omega_y}\right) \mathrm{sinc}(2\Omega_x x - m)\mathrm{sinc}(2\Omega_y y - n). \qquad (2.10)$$

Here the sinc function is defined as:

$$\mathrm{sinc}(x) = \frac{\sin(\pi x)}{\pi x}. \qquad (2.11)$$

A band-limited function is, therefore, typically expected to be well represented by $(4\Omega_x\Omega_y)(4L_xL_y)$ samples in an interval of size $(2L_x \times 2L_y)$. This number is, therefore, called the space-bandwidth product or degrees of freedom in a band-limited signal. For two-dimensional signals or images, the space-bandwidth product may be considered to be the same as the number of pixels required to represent the image. The sampling expansion is an orthogonal series representation of the band-limited signal, since the shifted sinc functions have the following orthogonality property:

$$\int_{-\infty}^{\infty} dx\, \mathrm{sinc}(2\Omega_x x - m)\, \mathrm{sinc}(2\Omega_x x - n) = \frac{1}{2\Omega_x}\delta_{m,n}. \qquad (2.12)$$

Here $m$ and $n$ are integers and $\delta_{m,n}$ is the Kronecker delta symbol which is equal to 1 for $m = n$ and zero otherwise. In optical systems the wave-field functions are typically band-limited due to the system aperture which limits the spatial frequencies passed by the system.

## 2.2 Review of random processes theory

The study of light propagation through turbulence requires us to deal with the phenomenon of random time-varying fluctuations in the local refractive index of air. The properties of a well defined beam propagating in vacuum are essentially deterministic. For a given initial amplitude, phase, and polarization profiles of a spatially coherent beam, its characteristics after propagation in vacuum over certain distance can be readily estimated. The refractive index of atmosphere is close to that of vacuum but is not a constant. Even small random fluctuations in the local refractive index of the order of $10^{-5}$–$10^{-6}$ can significantly disturb the beam characteristics after it has propagated over a distance of a few kilometers. Further, if the light beam under study is partially coherent, the $E$-field is not deterministic and shows inherent space–time fluctuations that need to be treated as a random process. Overall the ideas of the random process theory are therefore important for us throughout this book.

The numerical values assumed by a variable $u$ representing a physical quantity as a function of some continuous parameter $t$ (e.g. time) may not be predictable. They may, however, follow a probability distribution function $p(u, t)$. At a given $t$, integrating over all possible values that may be taken by $u$, we may write:

$$\int_{-\infty}^{\infty} du \, p(u, t) = 1. \tag{2.13}$$

We may imagine an ensemble of processes $u^{(1)}(t)$, $u^{(2)}(t)$, ... , $u^{(M)}(t)$ derived out of the same probability distribution $p(u, t)$. Any theoretical treatment of the quantity $u$ and description of what is typically observed in an experiment will necessarily have to be in terms of averages over this ensemble. Any conclusions based on the behavior of individual realizations of the process may not be meaningful. The expected mean of $u$ for example may be evaluated as:

$$\langle u \rangle = \int_{-\infty}^{\infty} du \, p(u, t) u. \tag{2.14}$$

The notation $\langle \ldots \rangle$ above denotes the ensemble average. While the probability density $p(u, t)$ provides statistical information about $u$ at $t$, additional interesting information about $u$ can be obtained by knowing about its correlations at $t_1$ and $t_2$. This correlation information is contained in the joint probability density $p(u_1, t_1; u_2, t_2)$ which is proportional to the probability that $u$ takes the values $u_1$ at $t_1$ and $u_2$ at $t_2$. In principle, joint densities of all higher orders are required for the complete description of the random process. In practice the most important correlation used is of second order and is denoted as:

$$\Gamma(t_1, t_2) = \langle u^*(t_1) \, u(t_2) \rangle. \tag{2.15}$$

The complex conjugate on the first term above makes the definition Hermitian symmetric:

$$\Gamma(t_1, t_2) = \Gamma^*(t_2, t_1). \tag{2.16}$$

Also when $t_1 = t_2$, this quantity simply reduces to the expected intensity $\langle |u|^2 \rangle$ of the underlying field. A random process $u(t)$ is called wide sense stationary if both $p(u, t)$ and the joint probability density $p(u_1, t_1; u_2, t_2)$ have no preferred origin in time $t$. The correlation function $\Gamma(t_1, t_2)$ in such cases is only a function of the time difference $(t_2 - t_1)$. A random process is called ergodic if the expectation values associated with the process can be evaluated using a long time average over its single realization. For example, for an ergodic process $u(t)$,

$$\langle u \rangle = \lim_{T \to \infty} \frac{1}{T} \int_{-T/2}^{T/2} dt \, u^{(n)}(t). \tag{2.17}$$

A sufficient condition for ergodicity to hold is that the correlation function $\Gamma(\tau = t_2 - t_1)$ dies out to zero for large $\tau$. In this case a single realization of the process can be divided into uncorrelated intervals that may themselves be considered as separate realizations of the random process. An important result related to the correlation function that we will have occasion to use is the Wiener–Khintchine theorem [3, 4] which states that the spectral density or spectrum of the random process is related to the correlation function by a Fourier transform relation:

$$S(\nu) = \int_{-\infty}^{\infty} d\tau \, \Gamma(\tau) \exp(-i2\pi\nu\tau).$$ (2.18)

We define the truncated Fourier transform of the random process $u(t)$ as:

$$U(\nu, T) = \int_{-T/2}^{T/2} dt \, u(t) \exp(-i2\pi\nu t).$$ (2.19)

The result in equation (2.18) can be proved by ensemble averaging the periodogram $S_T(\nu)$ defined as

$$S_T(\nu) = \frac{|U(\nu, T)|^2}{T},$$ (2.20)

in the long time limit. In other words, it can be shown that

$$S(\nu) = \lim_{T \to \infty} \left\langle \frac{|U(\nu, T)|^2}{T} \right\rangle.$$ (2.21)

It is easy to understand the Wiener–Khintchine relation intuitively. If we associate $u(t)$ with the $E$-field of light, it is well-known for example that for a narrow-band source, the correlation $\Gamma(\tau)$ survives longer, whereas for a broadband source the correlation dies down fast. In fact the extent over which $|\Gamma(\tau)|$ remains significant is termed the coherence time $\tau_c$. In describing refractive index fluctuations in the atmosphere, we will be concerned with a space-domain random process $n(\mathbf{r})$, which represents the refractive index of the atmosphere at location $\mathbf{r}$. If the correlation function of the process defined as:

$$\Gamma_n(\mathbf{r}_1, \mathbf{r}_2) = \langle n(\mathbf{r}_1) \, n(\mathbf{r}_2) \rangle,$$ (2.22)

is dependent only on the difference $(\mathbf{r}_1 - \mathbf{r}_2)$, then such a process is called homogeneous (analogous to the stationary process). In that case a Wiener–Khintchine spectrum for $\Gamma_n(\mathbf{r}_1 - \mathbf{r}_2)$ may be defined in a manner similar to equation (2.18), which plays an important role in the modeling of atmospheric turbulence as we shall see later in this book.

## 2.3 Simulating a random process with known spectral density

The nominal method to numerically generate a realization of random process $u(t)$ with a given spectral density $S(\nu)$, is the Fourier transform method. A random realization $u^{(k)}(t)$ for example can be generated by taking inverse Fourier transform of the quantity $\sqrt{S(\nu)} \times \exp(i\theta^{(k)}(\nu))$. Here $\theta^{(k)}(\nu)$ is a uniform or normally distributed random phase map. We observe that the ensemble averaged periodogram corresponding to $u^{(k)}(t)$ now automatically yields the desired spectral density $S(\nu)$ by design. As an illustration, we show three realizations of a random process

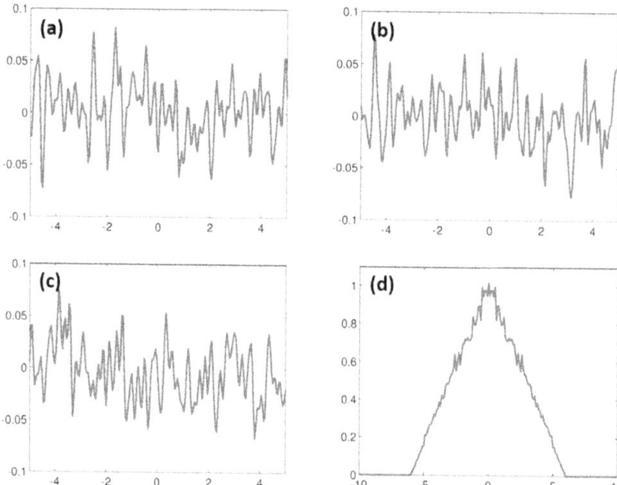

**Figure 2.1.** (a)–(c) Realizations of a random process with power spectrum $S(\nu)$ given by a triangle function in equation (2.23) with $\Delta\nu = 6$. The real part of the random process is shown. (d) Average of the Fourier transform magnitude squares calculated over 500 realizations of the random process. The $y$-axis units in (d) are arbitrary. The approximate triangular shape of the power spectrum is clearly observed.

described by the triangular spectrum in figures 2.1(a)–(c). Here the spectral density is defined as:

$$S(\nu) = \Lambda\left(\frac{\nu}{\Delta\nu}\right),\tag{2.23}$$

with $\Lambda(x) = 1 - |x|$ for $|x| \leqslant 1$ and equal to zero everywhere else. For the illustrations we have used the numerical value $\Delta\nu = 6$. Figure 2.1(d) shows the average Fourier power spectrum computed over 500 realizations of the random process. We observe that the shape of the average spectrum resembles that of the triangle function as expected.

## 2.4 Complex signal representation

Complex representation of real valued random processes is commonly used in the description of optical fields, coherence theory and in optical signal processing. We will examine this concept in detail in this section. The Fourier transform $U(\nu)$ of a real valued signal $u_r(t)$ has the symmetry property:

$$U(\nu) = U^*(-\nu).\tag{2.24}$$

As a result the negative frequency components do not contain any additional information compared to the positive frequency components. The negative frequencies may, therefore, be omitted without any loss of information. This may be achieved by multiplying the Fourier transform $U(\nu)$ with a unit step function. The corresponding time domain signal is referred to as the complex or analytic signal representation as first suggested by Gabor. The complex signal $u_z(t)$ can therefore be represented as:

$$u_z(t) = \frac{1}{2} \int_{-\infty}^{\infty} d\nu\; U(\nu)\,[1 + \mathrm{sgn}(\nu)]\exp(i2\pi\nu t).\qquad (2.25)$$

Here $\mathrm{sgn}(\nu)$ is the signum function defined as:

$$\begin{aligned}\mathrm{sgn}(\nu) &=\; 1 \;\; \text{for}\;\; \nu > 0\\ &= -1 \;\; \text{for}\;\; \nu < 0.\end{aligned}\qquad (2.26)$$

The term $[1 + \mathrm{sgn}(\nu)]/2$ in the above expression represents the unit step function. The sgn filter in frequency domain, as in equation (2.25), corresponds to Hilbert transform in the real (or $t$) space and the complex signal can be expressed as:

$$u_z(t) = \frac{1}{2}u(t) + \frac{i}{2\pi}P\int_{-\infty}^{\infty} dt'\;\frac{u(t')}{t - t'}.\qquad (2.27)$$

The symbol $P$ here denotes the principal value of the integral. For a narrow-band optical field with central frequency $\nu_0$, the real valued field is typically represented as

$$u_r(t) = a(t)\cos[2\pi\nu_0 t + \Phi(t)],\qquad (2.28)$$

where, $a(t)$ and $\Phi(t)$ are amplitude and phase functions that vary slowly compared to the carrier-frequency signal at frequency $\nu_0$. If $u_r(t)$ represents the scalar field associated with realistic laboratory sources, $a(t)$ and $\Phi(t)$ are not deterministic and have to be treated as random processes. Using the relations in equations (2.25) and (2.27), it can be shown that:

$$\begin{aligned}u_z(t) &= \frac{1}{2}a(t)(\cos[2\pi\nu_0 t + \Phi(t)] + i\,\sin[2\pi\nu_0 t + \Phi(t)])\\ &= \frac{1}{2}a(t)\exp[i2\pi\nu_0 t + i\Phi(t)].\end{aligned}\qquad (2.29)$$

We observe that in equation (2.28), we are representing a single function $u_r(t)$ in terms of two functions $a(t)$ and $\Phi(t)$. Such a representation is not, therefore, unique and an additional criterion is required to make an appropriate choice of the two functions $a(t)$ and $\Phi(t)$. The complex representation above using the Hilbert transform relation in equation (2.27) was obtained by suppressing the redundant information. One may ask if this representation is optimal in some sense. This question was addressed by Mandel in an interesting early work [5] which for some reason is not widely known in the optics as well as signal processing community.

Mandel's result is as follows. Suppose $u_r(t)$ represents a real valued random process. We will consider another random process generated by convolving $u_r(t)$ with a filter $k(t)$:

$$u_r'(t) = \int_{-\infty}^{\infty} dt'\; k(t - t')\,u_r(t),\qquad (2.30)$$

and construct a complex signal:

$$v_z(t) = u_r(t) + iu_r'(t)$$
$$= u_{zo}(t)\exp(i2\pi\nu_0 t).$$

(2.31)

For a stationary random process $u_r(t)$, the power spectral density is $S_r(\nu)$ (defined as per the Wiener–Khintchine theorem equation (2.18)) and the power spectral density $S_z(\nu)$ for the process $v_z(t)$ may be related by:

$$S_z(\nu) = S_r(\nu)|1 + iK(\nu)|^2.$$

(2.32)

Mandel's approach was to seek an appropriate filter $k(t)$ such that the fluctuation in the complex envelope $u_{zo}(t)$ defined by:

$$(Du_z)^2 = \left\langle \left| \frac{du_{zo}(t)}{dt} \right|^2 \right\rangle,$$

(2.33)

is minimized. The solution of this minimization problem can be obtained analytically by methods of variational calculus. The minimization of the quantity $(Du_z)^2$ in equation (2.33) can be shown to be equivalent to the solution of the following constrained optimization problem:

$$\text{minimize} \int_{-\infty}^{\infty} d\nu(\nu - \nu_0)^2 S_z(\nu),$$

(2.34)

with the constraint that

$$\int_{-\infty}^{\infty} d\nu(\nu - \nu_0)S_z(\nu) = 0.$$

(2.35)

The solution of this minimization problem as shown by Mandel is that $k(t)$ is the Hilbert transform that occurs in the second term of equation (2.27). Since the Hilbert transform connects cosines and sines (in one dimension), Mandel's result tells us that the cosine–sine quadrature representation commonly used for optical fields is special as it provides the most efficient (or least fluctuating) representation for the complex envelope.

## 2.5 Spiral phase quadrature transform

The cosine–sine quadrature representation in one dimension and its specialty in the sense of Mandel's theorem leads us to seek similar representation in higher dimensions. In this section we will extend Mandel's idea to two-dimensional signals in order to determine a transform equivalent to the Hilber transform. The treatment presented here essentially follows the work described in [6]. Later in this book we will see that the notion of quadrature transform in two dimensions as presented here has important implications for designing laser beams that can maintain robust intensity profiles on passing through turbulence.

We consider a two-dimensional signal $g(r, \theta)$ which is a realization of a stationary (or homogeneous) random field. We will assume the mean $\langle g(r, \theta) \rangle$ to be zero. The stationarity implies that the correlation function $\Gamma(\mathbf{r_1}, \mathbf{r_2}) = \langle g^*(\mathbf{r_1})g(\mathbf{r_2})\rangle$ is only a function of the difference $(\mathbf{r_1} - \mathbf{r_2})$ and the power spectral density $S(\rho)$ is related to the correlation function $\Gamma(\mathbf{r_1} - \mathbf{r_2})$ via the Wiener–Khintchine theorem. The spectral density $S(\rho)$ of a real valued process $g(r)$ is expected to have inversion symmetry:

$$S(\rho, \phi) = S(\rho, \phi + \pi). \tag{2.36}$$

In analogy with the one-dimensional case, we seek a filter function $K(\rho, \phi)$ that will be a generalization of the $\mathrm{sgn}(\nu)$ filter corresponding to the Hilbert transform and define the complex signal as:

$$g_z(r, \theta) = \frac{1}{2} \int \int \rho \, d\rho \, d\phi \, G(\rho, \phi)[1 + iK(\rho, \phi)]\exp[i2\pi r\rho \cos(\theta - \phi)]. \tag{2.37}$$

The spectral density of $g_z(r, \theta)$ denoted by $S_z(\rho, \phi)$ can thus be written as:

$$S_z(\rho, \phi) = \frac{1}{4}S(\rho, \phi)|1 + i\, K(\rho, \phi)|^2. \tag{2.38}$$

Since the $\mathrm{sgn}(\nu)$ filter in one dimension preserves the magnitude of the signal at all non-zero frequencies, we will further look for a unit magnitude filter of the form:

$$K(\rho, \phi) = \exp[i\alpha(\rho, \phi)], \tag{2.39}$$

with the requirement

$$K(\rho, \phi + \pi) = -K(\rho, \phi), \tag{2.40}$$

so that, the two-dimensional definition of the quadrature transform remains consistent with the one-dimensional definition. Next we define a complex envelope $a(r, \theta)$ corresponding to the signal $g_z(r, \theta)$:

$$a(r, \theta) = \frac{1}{2} \int \int \rho \, d\rho \, d\phi \, G(\rho, \phi)[1 + i\, K(\rho, \phi)]$$
$$\times \exp[i2\pi r(\rho - \rho_0)\cos(\theta - \phi)]. \tag{2.41}$$

Here $\rho_0$ is the carrier-frequency of the signal $g_z(r, \theta)$ defined as:

$$\rho_0 = \frac{\int \int \rho d\rho \, d\phi \, [\rho \, S_z(\rho, \phi)]}{\int \int \rho d\rho \, d\phi \, [S_z(\rho, \phi)]}. \tag{2.42}$$

The fluctuation in the envelope function analogous to equation (2.33) for this two-dimensional case may be defined using the Parseval theorem:

$$\left\langle \left| \frac{\partial a(r, \theta)}{\partial r} \right|^2 + \frac{1}{r^2} \left| \frac{\partial a(r, \theta)}{\partial \theta} \right|^2 \right\rangle = \pi^2 \int \int \rho \, d\rho \, d\phi \, (\rho - \rho_0)^2 S(\rho, \phi)$$
$$\times |1 + iK(\rho, \phi)|^2$$
$$= 4\pi^2 \int \int \rho \, d\rho \, d\phi \, (\rho - \rho_0)^2 S_z(\rho, \phi). \tag{2.43}$$

The task of finding the appropriate filter function thus reduces to the following variational optimization problem:

$$\text{minimize} \int \int \rho \, d\rho \, d\phi \, (\rho - \rho_0)^2 \, S_z(\rho, \phi),\tag{2.44}$$

subject to the constraint:

$$\int \int \rho \, d\rho \, d\phi \, (\rho - \rho_0) \, S_z(\rho, \phi) = 0.\tag{2.45}$$

The variational optimization problem is, therefore, essentially the same as that described in equations (2.34) and (2.35) respectively for the one-dimensional case. Note that we have assumed the filter $K(\rho, \phi)$ to be in the form $\exp[i\alpha(\rho, \phi)]$, so that,

$$|1 + iK(\rho, \phi)|^2 = 2[1 - \sin[\alpha(\rho, \phi)]].\tag{2.46}$$

Assuming that the solution to the minimization problem in equation (2.44) to be $\alpha(\rho, \phi)$, if we vary the solution to the neighboring function $\alpha(\rho, \phi) + \varepsilon\eta(\rho, \phi)$ for small positive number $\varepsilon$, then the integral in equation (2.44) must remain unchanged to first order in $\varepsilon$. In performing this variation the function $\eta(\rho, \phi)$ only needs to satisfy the condition in equation (2.40):

$$\exp[i\alpha(\rho, \phi) + i\varepsilon\,\eta(\rho, \phi)] = -\exp[i\alpha(\rho, \phi + \pi) + i\varepsilon\,\eta(\rho, \phi + \pi)].\tag{2.47}$$

which is already satisfied when $\varepsilon = 0$. Therefore, $\eta(\rho, \phi)$ is arbitrary but needs to satisfy the symmetry condition:

$$\eta(\rho, \phi) = \eta(\rho, \phi + \pi).\tag{2.48}$$

The variation in $\alpha(\rho, \phi)$ also changes the mean frequency $\rho_0$ to $(\rho_0 - \delta\rho_0)$ with $\delta\rho_0$ proportional to $\varepsilon$. Making these substitutions in equation (2.44) we can collect the terms that are first order in $\varepsilon$ and set their sum to zero.

$$2\delta\rho_0 \int \int \rho \, d\rho \, d\phi \, (\rho - \rho_0)S(\rho, \phi)\{1 - \sin[\alpha(\rho, \phi)]\}$$
$$- \varepsilon \int \int \rho \, d\rho \, d\phi \, (\rho - \rho_0)^2 \, S(\rho, \phi)\eta(\rho, \phi)\cos[\alpha(\rho, \phi)] = 0.\tag{2.49}$$

The first term above vanishes due to the constraint in equation (2.45). In view of the symmetry properties of $S(\rho, \phi)$, $K(\rho, \phi)$ and $\eta(\rho, \phi)$, the integral in the second term can be rewritten as:

$$\int_0^\pi d\phi \int \rho \, d\rho \, (\rho - \rho_0)^2 S(\rho, \phi)\eta(\rho, \phi)$$
$$\{\cos[\alpha(\rho, \phi)] + \cos[\alpha(\rho, \phi + \pi)]\} = 0.\tag{2.50}$$

In order for the above integral to vanish identically for arbitrary functions $S(\rho, \phi)$ and $\eta(\rho, \phi)$, we must have:

$$\cos[\alpha(\rho, \phi)] + \cos[\alpha(\rho, \phi + \pi)] = 0.\tag{2.51}$$

We observe that the simplest choice of such a function that satisfies this condition is:

$$\alpha(\rho, \phi) = \phi. \tag{2.52}$$

The filter corresponding to this simple choice is:

$$K(\rho, \phi) = \exp(i\phi). \tag{2.53}$$

The filter is thus defined everywhere except at $\rho = 0$ where $\phi$ is not defined. This should not be a problem for us since $g(r, \theta)$ is assumed to be a zero mean process which has a zero dc component. We therefore have an important result that the spiral phase filter $K(\rho, \phi) = \exp(i\phi)$ defined for two-dimensional signals is analogous to the sgn($\nu$) filter in one dimension in the sense of Mandel's theorem. The spiral phase filtering operation can thus be thought of as the quadrature transform for two-dimensional signals. The work of Larkin and co-workers [7] has already suggested through a number of numerical experiments that the spiral phase filter indeed is a suitable extension of the concept of Hilbert transform to two-dimensional signals. It is important to note that the Bessel functions of the first kind $J_0(ar)$ and $J_1(ar)$ with $r = \sqrt{x^2 + y^2}$ are exactly related by the spiral phase transform apart from some constant multipliers.

As an illustration of the spiral phase transform we consider the two-dimensional low frequency band by circle function:

$$\mathrm{circ}\left(\frac{\rho}{a}\right) = 1 \quad \text{for} \quad \rho \leqslant a \tag{2.54}$$
$$= 0 \quad \text{otherwise.}$$

Here $\rho = \sqrt{f_x^2 + f_y^2}$ is the radial coordinate in the two-dimensional Fourier space. The two functions

$$g_1(r, \theta) = \mathcal{F}^{-1}[\mathrm{circ}(\rho/a)], \tag{2.55}$$

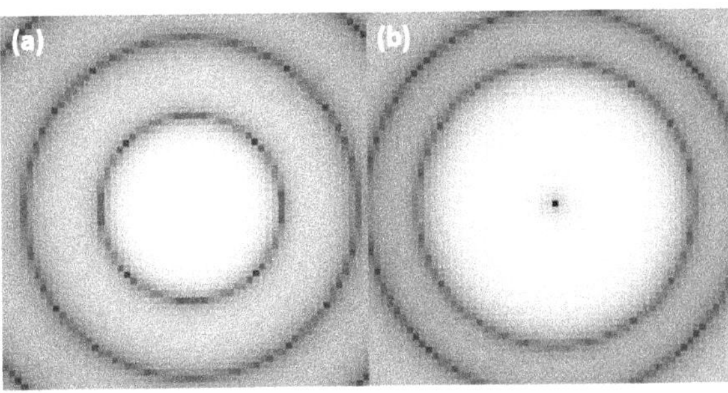

**Figure 2.2.** Illustration of complementary signals produced by the spiral phase transform. (a) and (b) show central portions (57 × 57 pixels) of $|g_1(r, \theta)|$ and $|g_2(r, \theta)|$ as defined in equations (2.55) and (2.56) respectively. The complementarity in the magnitude patterns can be seen visually.

and

$$g_2(r, \theta) = \mathcal{F}^{-1}[\exp(i\phi)\ \mathrm{circ}(\rho/a)], \qquad (2.56)$$

with $\phi = \arctan(f_y/f_x)$ should therefore form a 'two-dimensional quadrature' pair of functions showing sine–cosine like diversity. For numerical computation, we define the circ function over a $256 \times 256$ pixel array and choose $a = 8$. The central $57 \times 57$ portions of the Fourier transform magnitude images associated with $|g_1(r, \theta)|$ and $|g_2(r, \theta)|$ defined above are shown in figure 2.2. We observe that the locations of maxima and minima in the two images in figure 2.2 are indeed exchanged suggesting sine–cosine like diversity in the functions $|g_1(r, \theta)|$ and $|g_2(r, \theta)|$.

The spiral phase transform is closely related to the orbital angular momentum states of light that are central to the discussion in this book. We will show later that the complementarity above holds for general amplitude-phase apertures. In this context, the result will also be found to be useful when we discuss the engineering of laser beams that show robust intensity profiles on propagation through turbulence.

# References


[1] Bracewell R N 1987 *The Fourier Transform and Its Applications* (New York: McGraw-Hill)

[2] Osgood B G 2020 *EE261: The Fourier Transform and Its Applications* Stanford Engineering Online Course Material https://see.stanford.edu/course/ee261

[3] Mandel L and Wolf E 1995 *Optical Coherence and Quantum Optics* (Cambridge: Cambridge University Press)

[4] Goodman J W 2015 *Statistical Optics* 2nd edn (New York: Wiley)

[5] Mandel L 1967 Complex representation of optical fields in coherence theory *J. Opt. Soc. Am.* **57** 613–7

[6] Khare K 2008 Complex signal representation, Mandel's theorem, and spiral phase quadrature transform *Appl. Opt.* **47** E8–12

[7] Bone D J, Larkin K G and Oldfield M A 2001 Natural demodulation of two-dimensional fringe patterns. I. general background of the spiral phase quadrature transform *J. Opt. Soc. Am. A* **18** 1862–70

**IOP** Publishing

## Orbital Angular Momentum States of Light
Propagation through atmospheric turbulence
**Kedar Khare, Priyanka Lochab and Paramasivam Senthilkumaran**

# Chapter 3

# The angular spectrum method

Propagation of wave-fields is of basic importance to this book and we will discuss the angular spectrum approach [1] in this context. While our main concern is propagation of light through turbulence, it is important to lay out the methodology for propagation of fields in free space. Later we will see that the typical methodology for modeling atmospheric propagation involves propagating wave-fields through free space interspersed with random phase screens, where the angular spectrum method will be useful.

## 3.1 Wave equation

The wave equation is obtained from the Maxwell's equations and describes the propagation of light through a medium. For a general medium the Maxwell equations are given by:

$$\nabla \times \boldsymbol{H} = \frac{\partial \boldsymbol{D}}{\partial t} + \boldsymbol{J}, \tag{3.1}$$

$$\nabla \times \boldsymbol{E} = -\frac{\partial \boldsymbol{B}}{\partial t}, \tag{3.2}$$

$$\nabla \cdot \boldsymbol{D} = \rho, \tag{3.3}$$

$$\nabla \cdot \boldsymbol{B} = 0. \tag{3.4}$$

Here $\boldsymbol{H}$, $\boldsymbol{E}$, $\boldsymbol{D}$ and $\boldsymbol{B}$ represent magnetic field, electric field, electric displacement and the magnetic induction respectively. The symbol $\rho$ stands for the volume free charge density and $\boldsymbol{J}$ gives the electric current density. The magnetic field $\boldsymbol{H}$ and the electric displacement $\boldsymbol{D}$ are related to $\boldsymbol{B}$ and $\boldsymbol{E}$ through the constitutive relations. In a linear, isotropic medium $\boldsymbol{H}$ and $\boldsymbol{D}$ are related to $\boldsymbol{B}$ and $\boldsymbol{E}$ as:

doi:10.1088/978-0-7503-2280-5ch3

$$H = \frac{B}{\mu}, \tag{3.5}$$

$$D = \varepsilon E, \tag{3.6}$$

where $\mu$ is permeability and $\varepsilon$ is the permittivity of the medium. For free space, the permeability and dielectric constant have values

$$\mu_o = 4\pi \times 10^{-7} \ \mathrm{H} \ \mathrm{m}^{-1} \tag{3.7}$$

$$\varepsilon_o = 8.854 \times 10^{-12} \ \mathrm{F} \ \mathrm{m}^{-1} \tag{3.8}$$

and the speed of light in free space is given by:

$$c = \frac{1}{\sqrt{\mu_o \varepsilon_o}} = 3 \times 10^8 \ \mathrm{m} \ \mathrm{s}^{-1}. \tag{3.9}$$

The permeability and permittivity of a medium are related to their vacuum values through the relations:

$$\mu = \mu_r \mu_o, \tag{3.10}$$

$$\varepsilon = \varepsilon_r \varepsilon_o, \tag{3.11}$$

where $\mu_r$ and $\varepsilon_r$ are known as the relative permeability and relative permittivity of the medium, respectively. The dielectric constant can also be written in terms of the refractive index $n$ as:

$$\varepsilon = \varepsilon_o n^2. \tag{3.12}$$

Let us assume that the medium has constant magnetic permeability and zero conductivity. Taking the curl of equation (3.2) and rearranging the partial derivative and curl operator on the right-hand side, we get:

$$\nabla \times (\nabla \times E) = -\frac{\partial}{\partial t}(\nabla \times B). \tag{3.13}$$

The double curl operation on the left-hand side can be simplified using the vector identity:

$$\nabla \times (\nabla \times E) = \nabla(\nabla \cdot E) - \nabla^2 E. \tag{3.14}$$

Substituting $H$ and $D$ in terms of $B$ and $E$ in equations (3.1) and (3.3) leads to:

$$\nabla \times B = \mu \frac{\partial}{\partial t}(\varepsilon E), \tag{3.15}$$

$$\nabla \cdot (\varepsilon E) = 0. \tag{3.16}$$

We further evaluate equation (3.16) by using the vector identity:

$$\nabla \cdot (uA) = u(\nabla \cdot A) + A \cdot (\nabla u),$$

which gives us;

$$\nabla \cdot (\varepsilon E) = \varepsilon(\nabla \cdot E) + E \cdot (\nabla \varepsilon) = 0, \tag{3.17}$$

or,

$$\nabla \cdot E = -E \cdot \frac{(\nabla \varepsilon)}{\varepsilon} = -E \cdot \nabla \ln \varepsilon. \tag{3.18}$$

This allows us to write equation (3.13) in the form:

$$\nabla^2 E + \nabla(E \cdot \nabla \ln \varepsilon) = \mu \frac{\partial^2}{\partial t^2}(\varepsilon E). \tag{3.19}$$

The second term on the left-hand side gives the *depolarization of light* and represents the coupling between the orthogonal polarizations of the electric field on propagation through a medium which has space-dependent permittivity [2]. For a medium with constant $\varepsilon$, for example free space, the term $\nabla(E \cdot \nabla \ln \varepsilon) = 0$. We can then also write

$$\mu \frac{\partial^2}{\partial t^2}(\varepsilon E) = \mu \varepsilon \frac{\partial^2 E}{\partial t^2}.$$

These simplifications along with equations (3.9)–(3.12) give us the final form of the wave propagation equation in a medium of refractive index $n$ as:

$$\nabla^2 E(x, y, z, t) - \frac{n^2}{c^2}\frac{\partial^2}{\partial t^2}E(x, y, z, t) = 0. \tag{3.20}$$

A similar wave equation can also be written for the magnetic field $H$:

$$\nabla^2 H(x, y, z, t) - \frac{n^2}{c^2}\frac{\partial^2}{\partial t^2}H(x, y, z, t) = 0. \tag{3.21}$$

Since the same vector equation is satisfied by both $E$ and $H$, it means that all components of these vectors also satisfy an identical scalar equation. Therefore, we can for the moment ignore the vectorial nature of the wave equation and work with scalar quantities. Let the scalar optical field be given by $\tilde{U}(x, y, z, t)$, then

$$\nabla^2 \tilde{U}(x, y, z, t) - \frac{n^2}{c^2}\frac{\partial^2}{\partial t^2}\tilde{U}(x, y, z, t) = 0. \tag{3.22}$$

For monochromatic waves, the time-dependence of the optical fields is sinusoidal of the form $\exp(-i\omega t)$ where $\omega$ is the frequency of the wave,

$$\tilde{U}(x, y, z, t) = U(x, y, z)\exp(-i\omega t). \tag{3.23}$$

Substituting this form of $\tilde{U}(x, y, z, t)$ into equation (3.22), one gets the time-independent form of the scalar wave equation, also known as the *Helmholtz wave equation*,

$$(\nabla^2 + k^2)U(x, y, z) = 0, \tag{3.24}$$

where $k = \omega/c = 2\pi/\lambda$ is the wavenumber and $\lambda$ is the wavelength.

## 3.2 The angular spectrum formalism

The problem of interest to us involves solving the monochromatic wave equation or the Helmholtz equation:

$$(\nabla^2 + k^2)\, u(x, y; z) = 0, \tag{3.25}$$

between two planes transverse to the nominal propagation direction $(+z)$. Given the wave-field $u(x, y; 0)$ in the $z = 0$ plane, we would like to determine the field in the right half space. The Helmholtz equation in free space may be readily analyzed using a Fourier transform based method as we will discuss here. In later chapters of this book we will present the more involved case when the medium of propagation is filled with turbulent air with space–time varying refractive index. We begin by representing the field $u(x, y; z)$ as a two-dimensional Fourier expansion with respect to the transverse coordinates $(x, y)$:

$$u(x, y; z) = \int \int df_x\, df_y\ U(f_x, f_y; z) \exp[i2\pi(f_x x + f_y y)]. \tag{3.26}$$

Applying the operator $(\nabla^2 + k^2)$ on both sides of the above equation gives:

$$\int \int df_x df_y \left\{ \frac{\partial^2}{\partial z^2} + \left[ k^2 - 4\pi^2\!\left( f_x^2 + f_y^2 \right) \right] \right\} U(f_x, f_y; z)$$
$$\exp[i2\pi(f_x x + f_y y)] = 0. \tag{3.27}$$

In the above equation, we have used the derivative identity for the Fourier transform:

$$\frac{\partial u}{\partial x} = \int \int df_x\, df_y\ (2\pi i f_x) U(f_x, f_y; z) \exp[i2\pi(f_x x + f_y y)]. \tag{3.28}$$

A similar relation can be written for the $y$ derivative. The $z$-coordinate is being treated separately from the transverse coordinates as $z$ is the nominal direction in which we wish to propagate the field. Since the right-hand side of equation (3.27) is identically zero for an arbitrary function $U(f_x, f_y, z)$, the only way in which the integral can be zero is if the integrand itself is equal to zero. Therefore we have:

$$\left[ \frac{\partial^2}{\partial z^2} + \alpha^2 \right] U(f_x, f_y; z) = 0, \tag{3.29}$$

where we have introduced the notation:

$$\alpha^2 = k^2 - 4\pi^2\!\left( f_x^2 + f_y^2 \right). \tag{3.30}$$

The solution of equation (3.29) can be readily written as:

$$U(f_x, f_y, z) = A(f_x, f_y)\exp(i\alpha z) + B(f_x, f_y)\exp(-i\alpha z). \qquad (3.31)$$

We note that in the $\exp(-i2\pi\nu t)$ convention the first and second terms above represent the outgoing and the incoming solutions respectively. We further note that the spatial frequency plane may be divided into two distinct regions as shown in figure 3.1

$$f_x^2 + f_y^2 \leqslant \frac{1}{\lambda^2}, \qquad (3.32)$$

where $\alpha^2 \geqslant 0$ and the region

$$f_x^2 + f_y^2 > \frac{1}{\lambda^2}, \qquad (3.33)$$

where $\alpha$ is an imaginary quantity. For spatial frequencies outside the circle of radius $1/\lambda$ centered on origin of the two-dimensional Fourier space, the first term in equation (3.31) therefore decays exponentially whereas the second term grows exponentially. Since we are interested in the outgoing solution which needs to satisfy the boundary conditions at $z \to \infty$, we set $B(f_x, f_y) = 0$. Further, since the field in the $z = 0$ plane is already known, on setting $B(f_x, f_y) = 0$, it is easy to see that

$$A(f_x, f_y) = U(f_x, f_y; 0). \qquad (3.34)$$

The two-dimensional Fourier transform of the diffracted field after propagation by distance $z$ can therefore be represented as:

$$U(f_x, f_y; z) = U(f_x, f_y; 0)\exp(i\alpha z). \qquad (3.35)$$

Here it is to be noted that for the spatial frequencies $(f_x, f_y)$ satisfying the condition in equation (3.32), the quantity $\exp(i\alpha z)$ is purely a phase factor. The spatial frequencies satisfying the condition in equation (3.33) are seen to decay exponentially with $z$. The spatial frequency components within and outside the circle of

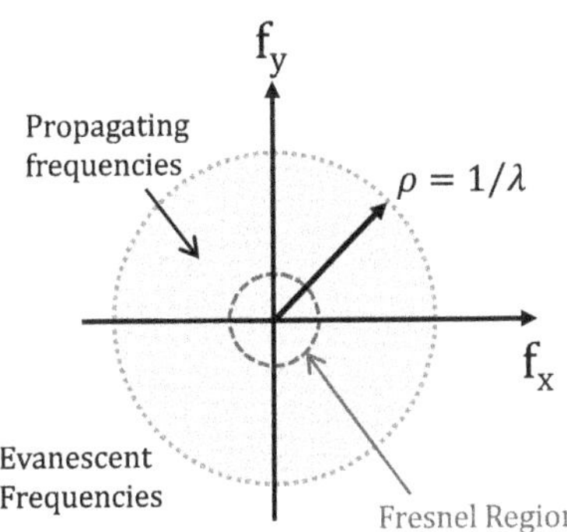


Figure 3.1. Division of spatial frequency plane into propagating and evanescent components. The Fresnel zone corresponds to paraxial frequency components close to the origin of the spatial frequency plane.

radius $1/\lambda$ in the two-dimensional Fourier space are, therefore, called propagating and evanescent components respectively [3, 4]. In this book, we will mostly not have occasion to deal with the evanescent components as the propagation distances of interest are in kilometers. The field $u(x, y, z)$ can be evaluated by inverse transforming the Fourier space relation in equation (3.35):

$$u(x, y, z) = \int \int df_x \, df_y \, [U(f_x, f_y; 0)\exp(i\alpha z)]\exp[i2\pi(f_x x + f_y y)]. \qquad (3.36)$$

We observe that the propagating part of the diffracted field can be decomposed as a linear combination of plane waves propagating in the directions represented by the vector $(f_x, f_y, \alpha/(2\pi))$ and having amplitudes $[U(f_x, f_y, 0)\exp(i\alpha z)]$. This methodology of treating the diffraction problem is therefore referred to as the angular spectrum approach. We observe that for the propagating components, the angle made by the $k$-vector of the plane wave with $x, y, z$ axes is $\theta_x = \arccos(\lambda f_x)$, $\theta_y = \arccos(\lambda f_y)$ and $\theta_z = \sqrt{1 - \cos^2(\theta_x) - \cos^2(\theta_y)}$ respectively. The low spatial frequency region near $(f_x, f_y) = (0, 0)$ thus corresponds to plane waves with wave-vector close to the $z$-axis (paraxial case). For the limiting spatial frequencies satisfying the relation $f_x^2 + f_y^2 = (1/\lambda^2)$, the corresponding plane waves have wave-vectors in the $x$-$y$ plane. The product relation in two-dimensional Fourier space as in equation (3.35) corresponds to a convolution in $(x, y)$ space and is given by the well-known first Rayleigh–Sommerfeld–Smythe diffraction formula [5]:

$$u(x, y; z) = \int \int dx' \, dy' \, u(x', y'; 0) \, \frac{\exp(ikR)}{2\pi R}\left(-ik + \frac{1}{R}\right)\frac{z}{R}. \qquad (3.37)$$

Here $R = \sqrt{(x - x')^2 + (y - y')^2 + z^2}$ is the distance between the points $(x', y', 0)$ and $(x, y, z)$. The relation in equation (3.37) is consistent with the Maxwell equations. It is easy to show that this relation reduces to the Fresnel and Fraunhofer relations in the appropriate paraxial limits. For example for optical wavelengths and typical laboratory distances $|-ik| \gg 1/R$ and for paraxial angles $z/R \approx 1$. This leads to the wide angle form of Fresnel diffraction:

$$u(x, y; z) = -ik \int \int dx' \, dy' \, u(x', y'; 0) \, \frac{\exp(ikR)}{2\pi R}. \qquad (3.38)$$

When $R$ is further approximated assuming $z \gg (x - x')$ and $z \gg (y - y')$ and in particular when

$$z^3 \gg \frac{k}{8}[(x - x')^2 + (y - y')^2]_{\max}, \qquad (3.39)$$

the above expression reduces to the familiar paraxial form [3]:

$$u(x, y; z) = \frac{-ike^{ikz}}{2\pi R} \int \int dx' \, dy' \, u(x', y'; 0) \, e^{\frac{i\pi}{\lambda z}[(x-x')^2+(y-y')^2]}. \qquad (3.40)$$

Finally when $z \gg \frac{k(x'^2 + y'^2)_{max}}{2}$, the Fresnel approximation formula above can be simplified further leading to the Fraunhofer approximation [3]:

$$u(x, y; z) = \frac{-ike^{ikz}\, e^{\frac{i\pi}{\lambda z}(x^2 + y^2)}}{2\pi R} \int \int dx'\, dy'\, u(x', y'; 0)\, e^{-i\frac{2\pi}{\lambda z}(xx' + yy')}. \tag{3.41}$$

Note that the Fraunhofer diffraction formula above shows that the far-field diffraction pattern for a finite aperture illuminated by a plane wave is nothing but the two-dimensional Fourier transform of the field in the $z = 0$ plane.

## 3.3 Sampling considerations and usage of fast Fourier transform routines

For general beam profile structures, it is easier to implement the free-space propagation calculation using the relation equation (3.35). The required forward and inverse Fourier transforms can be evaluated using the fast Fourier transform (FFT) routines that are readily available with standard programming tools. While propagating the field over a distance $z$ it is important to pay attention to sampling considerations. If the computational window size is $L \times L$, then over a distance $z$ the spatial frequencies involved are in the range:

$$\Delta f = \pm\frac{\sin\theta_{max}}{\lambda} = \pm\frac{(L/2)}{\lambda\sqrt{z^2 + (L/2)^2}}. \tag{3.42}$$

Here $\theta_{max}$ is the angle made by a line joining the on-axis point in the $z = 0$ plane with the edge of the computational window located at the plane $z = L$. As a result, when sampling the field at discrete points $(m\Delta x, n\Delta y)$ for integers $m$ and $n$, the sampling intervals must be selected to satisfy the sampling criterion:

$$\Delta x, \Delta y \leqslant \frac{1}{2\Delta f}. \tag{3.43}$$

Numerical computation of field propagation can be performed easily using a Fourier transform based approach as is evident from equation (3.35). Here we describe some finer points associated with usage of the FFT routines readily available in popular computational programs such as MATLAB, Octave, Python, etc, for the purpose of field propagation calculations. We note that the transfer function relation allows us to write the field $u(x, y; z)$ as:

$$u(x, y; z) = \mathcal{F}^{-1}\{\mathcal{F}[u(x, y; 0)]\exp(i\alpha z)\}. \tag{3.44}$$

Suppose that the field in the $z = 0$ plane is sampled over a window of $L \times L$ and the appropriate sampling criterion as in equation (3.43) has been met. For a one-dimensional function $u(m\Delta x)$ with $N = (L)/(\Delta x)$ samples, the FFT routine leads to the computation:

$$U(n) = \sum_{m=0}^{N-1} u(m\Delta x)\exp(-i2\pi mn/N), \quad n = 0, 1, 2, \dots, (N-1). \qquad (3.45)$$

The standard function call: $U = fft(u)$ performs the above numerical computation and leads to a new vector $U$ with the same length $N$. The spatial frequencies associated with the elements of this vector $U$ are ordered as:

$$0, \frac{1}{L}, \frac{2}{L}, \dots, \frac{N-1}{L}.$$

Note that the first element of the vector $U$ is associated with the zero frequency. Secondly note that the zero phase factor is applied to the first element $u(m = 0)$ of the input function. In a two-dimensional version of the function call (the `fft2` function), the zero frequency will be located at the left corner of the result matrix. For the optical field applications it is more appropriate that the zero phase be applied to the on-axis element, for example in accordance with the Fraunhofer diffraction formula in equation (3.41). A simple trick to make sure we get numerical results consistent with what is expected in optics laboratory experiments is to use the function call `fft` as follows: `U = fftshift(fft(ifftshift(u)))`. The function *ifftshift* shifts the central 'on-axis' element of the vector $u$ above to the first place by swapping the two halves of the vector. The function `fftshift` on the other hand readjusts the result of `fft` function call such that the zero frequency component is brought back on-axis to the central element. The two functions `fftshift` and `ifftshift` are identical when $N$ is even and differs by a single element shift when $N$ is odd. The function call suitable for inverse FFT is similar and is given by: $u$ = `fftshift(ifft(ifftshift(U)))`. Note that here `fft` has been replaced by `ifft` (inverse FFT), however, the sequence of `fftshift` and `ifftshift` has remained the same. We would like to emphasize here that the use of `fftshift` or `ifftshift` functions is only required here in order to obtain amplitude and phase profiles of the field that are consistent with what is expected in a laboratory experiment. This discussion is important for practical implementation of field propagation, otherwise, it is possible to get unphysical computational results (particularly for the phase map of the resultant fields).

## 3.4 Numerical propagation of fields in free space

As an illustration, we show the free-space propagation of a vortex beam described in the $z = 0$ plane as:

$$u(r, \theta, = 0) = A\exp(-r^2/w_o^2)\exp(-i\theta), \qquad (3.46)$$

where $r = \sqrt{x^2 + y^2}$ and $\theta = \arctan(y/x)$. As we will explain later in the book, this profile corresponds to a vortex beam in the orbital angular momentum state $l = 1$. The term 'vortex' here is suggestive of the phase singularity at $r = 0$ where phase is undefined. The amplitude and phase profile of the beam in $z = 0$ plane is shown in figures 3.2(a) and (b) respectively. The beam waist parameter $w_0$ is taken to be $100\lambda$ and the computational window is a square of size $L = 1000\lambda$. The propagation

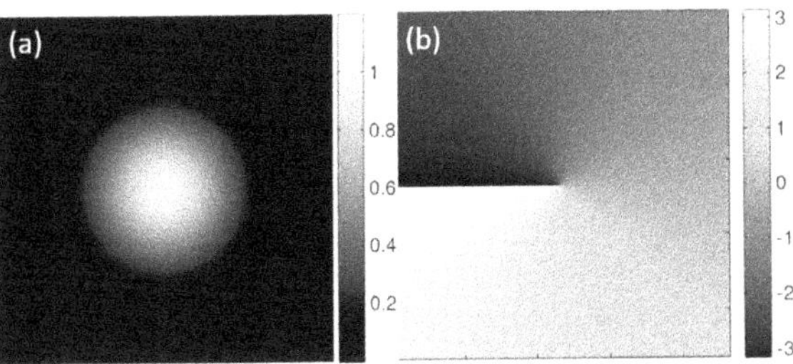

**Figure 3.2.** Amplitude and phase profile of the field described in equation (3.46) are shown in (a) and (b) respectively.

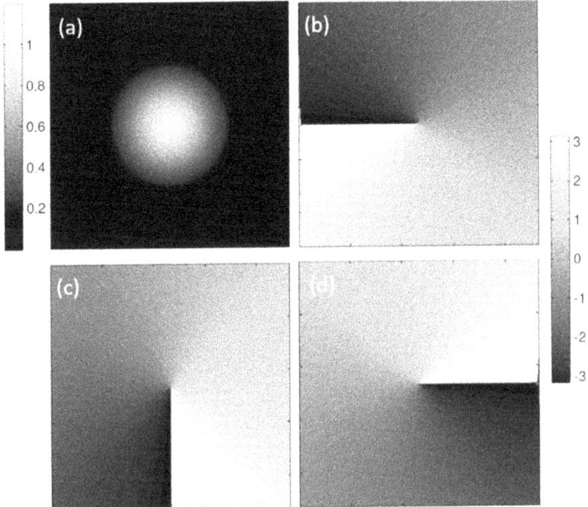

**Figure 3.3.** (a) Amplitude of the propagated field in figure 3.2 at $z = 100\lambda$, (b), (c), (d): phase map of the propagated field at distances $z = 100\lambda$, $100.25\lambda$, and $100.5\lambda$ respectively.

distance is set to $z = 100\lambda$, $100.25\lambda$, and $100.5\lambda$ respectively. For these propagation distances, the angular spectrum over the computational window is essentially purely propagating in nature. As per sampling criterion in equation (3.43), a square grid of $M \times M$ samples is required where $M = 929$. The amplitude of the field at $z = 100\lambda$ and the phase of the propagated field at the three distances are shown in figures 3.3(a), (b), (c) and (d) respectively. The amplitude of the field at $z = 100.25\lambda$ and $100.5\lambda$ is not shown separately as it is similar to that shown in figure 3.3(a). It is important to note that the phase jump at $\theta = \pi$ in the phase profile shows rotation by $90°$ and $180°$ for the distances $z = 100.25\lambda$ and $z = 100.5\lambda$ respectively. This is a well-known feature of a vortex beam that its phase jump nominally rotates with a periodicity of $\lambda$. This is an interesting feature of the field that has naturally come out from angular spectrum based propagation method.

While the numerical illustration shown here is approximately correct, we note that the input field in equation (3.46) changes very rapidly near the phase singularity at $r = 0$. The discrete sampling of the field at intervals $(\Delta x, \Delta y)$ is clearly not adequate to represent this fast varying field near the vortex core ($r = 0$) in this case. Examination of the field near the singularity needs additional discussion that will be provided in the next chapter. The periodicity ($=\lambda$) of the phase jump along the propagation distance $\lambda$ does not follow trivially from the Rayleigh–Sommerfeld diffraction formula and the discussion in the next chapter will show that this periodicity is generally observed only if we are not near the singularity at $r = 0$. These points however demand a more careful semi-analytical analysis using the angular spectrum approach.

# References


[1] Clemmow P C 1966 *The Plane Wave Spectrum Representation of Electromagnetic Fields* (Oxford: Pergamon)
[2] Tatarskii V I 1967 Depolarization of light by turbulent atmospheric inhomogeneities *Radiophys. Quantum Electron.* **10** 987–8
[3] Goodman J W 2016 *Introduction to Fourier Optics* 3rd edn (Boston, MA: Roberts)
[4] Khare K 2015 *Fourier Optics and Computational Imaging* 1st edn (New York: Wiley)
[5] Mandel L and Wolf E 1995 *Optical Coherence and Quantum Optics* (Cambridge: Cambridge University Press)

**IOP** Publishing

# Orbital Angular Momentum States of Light
Propagation through atmospheric turbulence

**Kedar Khare, Priyanka Lochab and Paramasivam Senthilkumaran**

# Chapter 4

# Near-core structure of a propagating optical vortex

## 4.1 Vortex propagation using the angular spectrum method

In this chapter we present an illustration of the angular spectrum method for a propagating optical vortex and observe some interesting features of the vortex field in the near-core region. Optical vortices are commonly pictured as having a helical wavefront with periodicity of the helix equal to the wavelength $\lambda$. However, this periodicity of $\lambda$ is somewhat puzzling as there seems to be no simple explanation for this periodicity purely based on free-space diffraction. Propagation of vortex fields in free-space thus demands additional investigation, as we describe in this chapter. The discussion in this chapter closely follows [1].

We begin by assuming an idealized initial scalar component of the wave-field in the $z = 0$ plane given by:

$$u(r, \theta; 0) = \exp(i\theta), \tag{4.1}$$

where $\theta = \arctan(y/x)$ and $r = \sqrt{x^2 + y^2}$. The angle $\theta$ is not defined at $r = 0$ and the field will therefore have a null at this position. It is important to note that the input field has a singularity at the origin and as $r \to 0$ the input field changes arbitrarily fast implying that the input field $u(r, \theta; 0)$ has high spatial frequency content. Since Fresnel or Fraunhofer diffraction approximations are limited to the paraxial regime, they cannot account for the high spatial frequencies in the input field. We therefore examine propagation of the helical wavefront using the angular spectrum approach. The two-dimensional Fourier transform of $u(r, \theta; 0) = \exp(i\theta)$ first needs to be evaluated in order to use the angular spectrum approach. The two-dimensional Fourier transform of $\exp(i\theta)$ may be evaluated in polar coordinates as:

$$\int_0^{2\pi} d\theta \int_0^{\infty} r \, dr \, \exp(i\theta)\exp[-i2\pi r\rho \cos(\theta - \phi)]$$
$$= -(2\pi i)\exp(i\phi)\int_0^{\infty} r \, dr J_1(2\pi r\rho). \tag{4.2}$$

Here $\phi = \arctan(f_y/f_x)$ is the polar angle in the two-dimensional spatial frequency plane and $\rho = \sqrt{f_x^2 + f_y^2}$. In the above equation we have evaluated the integral over $\theta$ using the integral representation for the Bessel function:

$$J_1(u) = \frac{1}{2\pi}\int_0^{2\pi} d\theta \, \exp(-i\theta)\exp(i\, u \sin\theta). \tag{4.3}$$

The integral of $r \, J_1(2\pi r\rho)$ on the right-hand side of equation (4.2) can be evaluated using integration by parts where it will be handy to use the Bessel function relations:

$$\frac{d}{dx} J_0(x) = -J_1(x), \tag{4.4}$$

and

$$\int_0^{\infty} du \, J_0(u) = 1. \tag{4.5}$$

The two-dimensional Fourier transform of $u(x, y, 0)$ may be obtained using these Bessel function relations and is given by:

$$U(f_x, f_y; 0) = \frac{-i \exp(i\phi)}{2\pi\rho^2}. \tag{4.6}$$

A general form of this result for arbitrary order vortex beam can be derived using properties of Bessel functions and is given by [2]:

$$\int_0^{2\pi} d\theta \int_0^{\infty} r \, dr \, \exp(in\theta)\exp[-i2\pi r\rho \cos(\theta - \phi)]$$
$$= \frac{|n|(-i)^{|n|}\exp(i\, n\, \phi)}{2\pi\rho^2}. \tag{4.7}$$

The field after propagation by distance $z$ as per the angular spectrum method is obtained by inverse Fourier transforming $U(f_x, f_y; z)$ and may be expressed as:

$$u(r, \theta; z) = \int\int \rho \, d\rho \, d\phi \, \frac{-i \exp(i\phi)}{2\pi\rho^2} \exp(i\, z\sqrt{k^2 - 4\pi^2\rho^2})$$
$$\times \exp[i2\pi r\rho \cos(\phi - \theta)], \tag{4.8}$$

where we have made use of equation (3.36) in polar coordinate form and equation (4.6). The integral over the polar angle $\phi$ can be readily evaluated giving:

$$u(r, \theta; z) = \exp(i\theta) \, V_1(r, z), \tag{4.9}$$

where

$$V_1(r, z) = \int_0^\infty d\rho \; \frac{\exp(iz\sqrt{k^2 - 4\pi^2\rho^2})}{\rho} J_1(2\pi r\rho). \qquad (4.10)$$

It may be noted that the $\rho$ in the denominator in the integrand in the above relation does not pose problem as $[J_1(2\pi r\rho)/\rho]$ is finite as $\rho \to 0$. From equation (4.9) we note an interesting result that on starting with the vortex wavefront $\exp(i\theta)$, the resultant wavefront at any distance $z$ is given by the same initial vortex function multiplied by $V_1(r, z)$ as in equation (4.10). It is important to note that the function $V_1(r, z)$ is identically equal to zero at $r = 0$ since the Bessel function in the integrand vanishes at this point. The phase of the helical wavefront may now be written as:

$$\Phi(r, \theta; z) = \arg[u(r, \theta; z)] = \theta + \arg[V_1(r, z)]. \qquad (4.11)$$

Contrary to the usual assumption about paraxial propagation of vortex wavefields, the phase of the helical wavefront is thus seen to depend on the radial coordinate $r$ as well. In the previous chapter (figure 3.3), this behavior of the propagating vortex field was not evident due to the lack of sufficient sampling of the field in the input plane. It is clear that at the origin ($r = 0$) both $\theta$ and $\arg[V_1(r, z)]$ are not defined and the propagated field continues to be null along the $z$-axis for arbitrary propagation distances.

## 4.2 Phase dip near the vortex core

In this section we will closely examine the near-core structure of a propagating optical vortex by numerically evaluating the results in equations (4.10) and (4.11). We have employed a standard numerical quadrature technique in order to evaluate the integral in the definition of $V_1(r, z)$. The lower integration limit is $\rho = 0$ and the upper limit used for numerical integration is $\rho = 10/\lambda$ thus including the evanescent components. Our computation showed that including the evanescent spatial frequency components beyond this range did not make any significant difference to the numerical results. In figure 4.1 we plot the phase of the function $V_1(r, z)$ for $r = 0.1\lambda$ and $r = 1\lambda$ as $z$ is varied from 0 to $1\lambda$. The curves in figure 4.1 are plotted for the cases when the upper limit of integration in equation (4.10) are set to $\rho = 0.5/\lambda$, $\rho = 1/\lambda$ and $\rho = 10/\lambda$. The plots in figure 4.1(a) show that as the limit of integration for $\rho$ is increased, the phase of $V_1(r, z)$ shows a dip right after $z = 0$. The black curve in this figure represents the numerical value of $kz$. The dip in phase significantly increases when evanescent frequencies are included in the computation compared to the case when only propagating components (up to $\rho = 1/\lambda$) are included. The contribution of evanescent components, therefore, seems to be important in this case in understanding the detailed near-core structure of a propagating vortex. The red curve representing the phase of $V_1(r, z)$ with inclusion of propagating as well as evanescent frequencies is seen to become parallel to the $kz$ line but always lags behind it due to the initial phase dip. The case of $r = 1\lambda$ as shown in figure 4.1(b) shows that all the curves as in figure 4.1(a) are now close the black line representing the quantity $kz$.

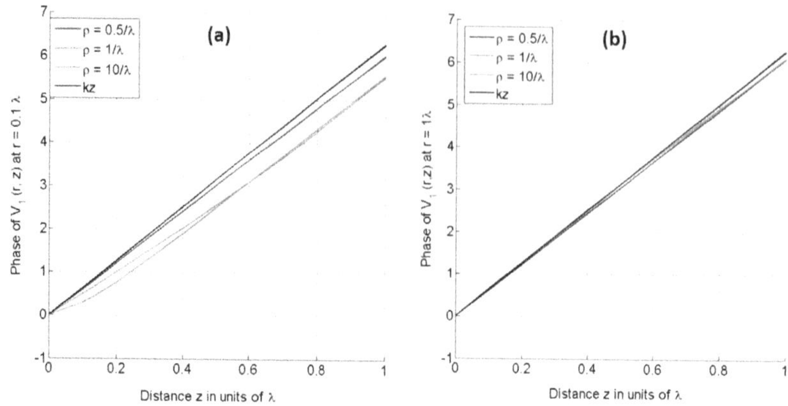

**Figure 4.1.** (a), (b): phase of function $V_1(r, z)$ near vortex core at $r = 0.1\lambda$ and $r = 1\lambda$ respectively. The $z$ variation is kept close to the input plane in order to distinguish the different curves clearly. Adapted with permission from [1] © The Optical Society.

In figures 4.2(a), (c), (e) we show the variation of phase of $V_1(r, z)$ as a function of $r$ at propagation distances of $z = 1\lambda$, $10\lambda$ and $100\lambda$ as two-dimensional surface plots. Figures 4.2(b), (d), (f) show the total phase $\Phi(r, \theta, z)$ for the corresponding distances. These results indicate that there is a phase dip near the core of a propagating vortex beam. The magnitude of the phase dip is close to $-0.8$ radians and the physical extent of the phase dip grows as the beam propagates from $z = 0$. Additional investigation is needed to understand magnitude of the phase dip near the core. In figures 4.2(b), (d), (f) the units on the $x$ and $y$ axes are in terms of wavelength $\lambda$. We observe that due to the phase dip the contours of equal phase are no longer radial near the vortex core ($r = 0$) but are seen to spiral around the core. The presence of phase lag suggests that a propagating vortex does not have a perfectly helical wavefront but near its core (the region near $r = 0$) the constant phase contours show a spiraling effect. This phase lag effect is not fully studied in the literature. We believe that the angular spectrum method is a good tool to examine such effects as it can naturally handle both propagating and evanescent components in the field. Any additional ideas such as super-oscillatory functions to describe the fast varying field near the core are then not required. As seen in figure 4.2, as the vortex beam propagates, the phase lag near-core continues to exist and the region with spiral lines of phase contours grows due to free-space propagation. For a propagating phase front it is common to associate the **k**-vector associated with the wavefront with the gradient $\nabla\Phi(r, \theta, z)$ of the wavefront phase function. Since the propagating vortex phase front has radial dependence, the transverse **k**-vector has a radial component near core in addition to the usual azimuthal component due to the first term in equation (4.11). The effect of this radial component of **k**-vector has been observed experimentally [3] in the form of an anomalous fringe period in Young's double slit experiment where two closely spaced slits are illuminated symmetrically by a vortex beam. In this case, the spiraling phase contour lines as observed in figure 4.2 lead to an additional wedge prism-like phase ramp at the two slits.

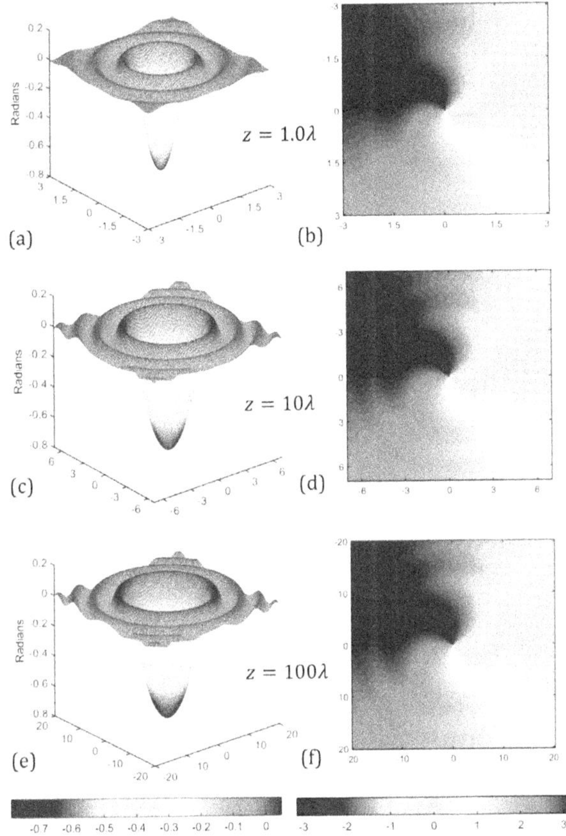

**Figure 4.2.** (a), (c) and (e) show the retarded part of the wavefront for different propagation distances while the corresponding total phase maps for $\Phi(r, \theta; z)$ are shown in (b), (d) and (f) respectively. The propagation distance $z$ for (a) and (b) is $z = 1\lambda$, (c) and (d) is $z = 10\lambda$ and (e) and (f) is $z = 100\lambda$. The range of $x$–$y$ coordinates in (a) and (b) go from $-3\lambda$ to $3\lambda$, in (c) and (d) it is from $-6.0\lambda$ to $6.0\lambda$, and lastly (e) and (f) go from $-20\lambda$ to $20\lambda$. Adapted with permission from [1] © The Optical Society

Depending on the location of the slits (near or far from core), this effect significantly changes the fringe period observed on a screen.

The detailed structure of higher order vortices may also be studied with the angular spectrum method to understand the rich structure of near-core fields. Using the result in equation (4.7), the vortex wavefront of the form $\exp(il\theta)$ can again be shown to be a product of two terms:

$$u(r, \theta, z) = \exp(il\theta)\, V_l(r, z), \tag{4.12}$$

where

$$V_l(r, z) = i^l l \exp(il3\pi/2) \int_0^\infty \frac{e^{iz\sqrt{k^2 - 4\pi^2\rho^2}}}{\rho} J_l(2\pi r\rho)\, d\rho. \tag{4.13}$$

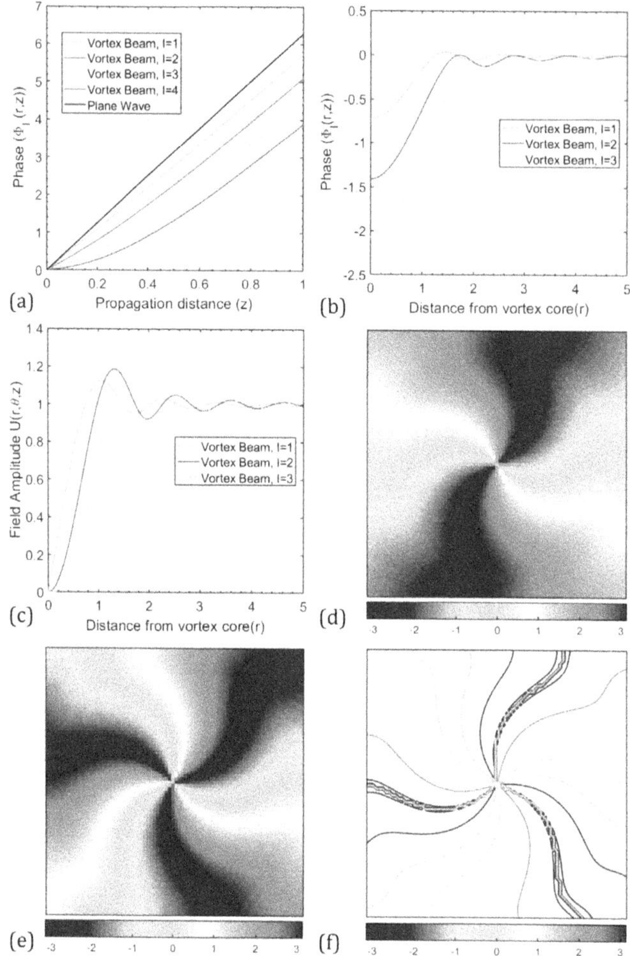

**Figure 4.3.** The variation of the phase function $\Phi(r, \theta, z)$ for different values of topological charge $l$ is shown. (a) $\Phi(r, \theta, z)$ (in radians) as a function of propagation distance $z$ at $r = 0.1\lambda$. (b) $\Phi(r, \theta, z)$ (in radians) as a function of the radial distance $r$ at $z = 1\lambda$. (c) Field amplitude $u(r, \theta, z)$ as a function of the radial distance $r$ at $z = 1\lambda$. (d) and (e) show the total phase map $\Phi(r, \theta, z)$ for $l = 2$ and $l = 3$ respectively at a propagation distance of $z = 1\lambda$. (f) shows the spiraling phase contour lines for $l = 3$ vortex. In the plots, the $z$ and $r$ values are given in units of $\lambda$. The range of $x$–$y$ coordinates in (d)–(f) is from $-2\lambda$ to $2\lambda$. Adapted with permission from [1] © The Optical Society.

In figure 4.3 we summarize the results for the total phase $\Phi(r, \theta, z)$ and the phase lag $V_l(r, z)$ for propagating vortices with charges 1, 2, and 3. From figure 4.3 we observe that the phase difference between the vortex beam and the plane wave ($kz - \Phi(r, z)$) increases with increasing the value of the topological charge. The size of the vortex core region is also seen to increase for higher $l$ values as seen from the field amplitude plot as shown in figure 4.3(c).

In summary this chapter used the angular spectrum formalism developed in chapter 3 to explore the nature of the near-core structure of optical vortices. Contrary to the usual paraxial treatment where helical wavefront phase functions of the form $\exp(il\theta)$ are associated with optical vortices, the angular spectrum picture suggests that there is a rich spiraling phase front structure near the core of optical vortices. The angular spectrum picture inherently incorporates all spatial frequencies and can therefore readily handle the fast varying wavefront field features in the near-core region.

# References

[1] Lochab P, Senthilkumaran P and Khare K 2016 Near-core structure of a propagating optical vortex *J. Opt. Soc. Am.* A **33** 2485–90

[2] Berry M V 2004 Optical vortices evolving from helicoidal integer and fractional phase steps *J. Opt. A: Pure Appl. Opt.* **6** 259–68

[3] Senthilkumaran P and Bahl M 2015 Young's experiment with waves near zeros *Opt. Express* **23** 10968–73

**IOP** Publishing

# Orbital Angular Momentum States of Light
## Propagation through atmospheric turbulence

**Kedar Khare, Priyanka Lochab and Paramasivam Senthilkumaran**

# Chapter 5

# Orbital angular momentum states of light

In the previous chapter we studied the free-space propagation of a vortex wavefront using the angular spectrum approach. It was observed that the helical phase fronts have rich structure near the core. The optical vortex beams represented by the azimuthal phase structure given by $\exp(il\theta)$ also have an interesting property that carries an intrinsic orbital angular momentum (OAM) of $l\hbar$ per photon. In this book we will refer to these states as the OAM states of light. This chapter provides an introduction to the scalar orbital angular momentum states and their generic properties. The azimuthal phase variation is accompanied by the presence of a phase singularity at the center of these beams. The amplitude of the helical wavefront at the on-axis point $r = 0$ vanishes and as explained later a line integral of phase around the zero amplitude point is equal to $(2l\pi)$. Their wavefront structure is helical or in some cases contains several intertwined helices depending on the value of integer $l$. The scalar phase singular beams when mixed in orthogonal polarizations can give rise to inhomogeneously polarized vector beams which host variety of polarization singular structures. In this chapter, we review the properties of scalar OAM beams. As a first task we will explain that the OAM phase structure naturally occurs in the paraxial solution of the wave equation in cylindrical $(r, \theta, z)$ coordinates [1–3].

## 5.1 Solutions of paraxial wave equation with phase singularities

If the propagation vectors of the optical waves subtend small angles with respect to the propagation axis, then they are called *paraxial waves* [4]. Assuming the nominal propagation direction to be along $z$-axis,

$$U(x, y, z) = u(x, y, z)\exp(ikz), \tag{5.1}$$

where $u(x, y, z)$ is a complex quantity and varies slowly with respect to $z$. The paraxial form of the wave equation can be obtained by substituting equation (5.1) into the Helmholtz equation and neglecting the term containing the second derivative of $u(x, y, z)$ with respect to $z$. Under the condition that

$$\lambda \left| \frac{\partial^2 u}{\partial z^2} \right| \ll \left| \frac{\partial u}{\partial z} \right|, \tag{5.2}$$

one can write the paraxial wave equation in the form:

$$\frac{\partial^2 u}{\partial x^2} + \frac{\partial^2 u}{\partial y^2} + 2ik\frac{\partial u}{\partial z} = 0. \tag{5.3}$$

In cylindrical coordinates, this equation can be written as:

$$\frac{1}{r}\frac{\partial}{\partial r}\left(r\frac{\partial u}{\partial r}\right) + \frac{1}{r^2}\frac{\partial^2 u}{\partial \theta^2} - 2ik\frac{\partial u}{\partial z} = 0. \tag{5.4}$$

With the separation of variables technique, the solutions of this equation are the Laguerre–Gaussian (LG) functions (or modes) [5] given by:

$$u(r, \theta, z) = \mathrm{LG}(p, l) = \sqrt{\frac{2^{|l|+1}p!}{\pi(|l| + p)!}} \frac{r^{|l|}\exp(il\theta)}{w_o^{|l|+1}(1 + iz/z_r)^{|l|+1}}\left(\frac{z_r - iz}{z_r + iz}\right)^p$$
$$\times \exp\left(\frac{-r^2}{w_o^2(1 + iz/z_r)}\right)L_p^{|l|}\left(\frac{2r^2}{w_o^2(1 + z^2/z_r^2)}\right). \tag{5.5}$$

Here, $w_o$ is the $1/e$ width of the Gaussian in the beam waist plane ($z = 0$), $z_r = kw_o^2/2$ is the Rayleigh range of the beam and $|l|$ and $p$ are the azimuthal and radial mode indices associated with the generalized Laguerre polynomial $L_p^{|l|}$,

$$L_p^{|l|}(x) = \frac{e^x x^{-|l|}}{p!}\frac{d^p}{dx^p}[e^{-x}x^{|l|+p}]. \tag{5.6}$$

The important quantity to note here is the integer $l$ which gives the topological charge of this beam through the term $\exp(il\theta)$. For $p = 0$, the two lowest order LG modes can be written in $z = 0$ plane as:

$$\mathrm{LG}(0, 0) = \sqrt{\frac{2}{\pi w_o^2}}\,\exp\left(-r^2/w_o^2\right), \tag{5.7}$$

$$\mathrm{LG}(0, 1) = \frac{2r}{\sqrt{\pi}\,w_o^2}\,\exp(i\theta)\exp\left(-r^2/w_o^2\right). \tag{5.8}$$

The LG(0,1) mode appears as a single annular intensity ring with $2\pi$ phase singularity along the beam axis (see figure 5.1(a)). A few other examples of the intensity and phase profile of different LG modes are shown in figures 5.1(b) and (c).

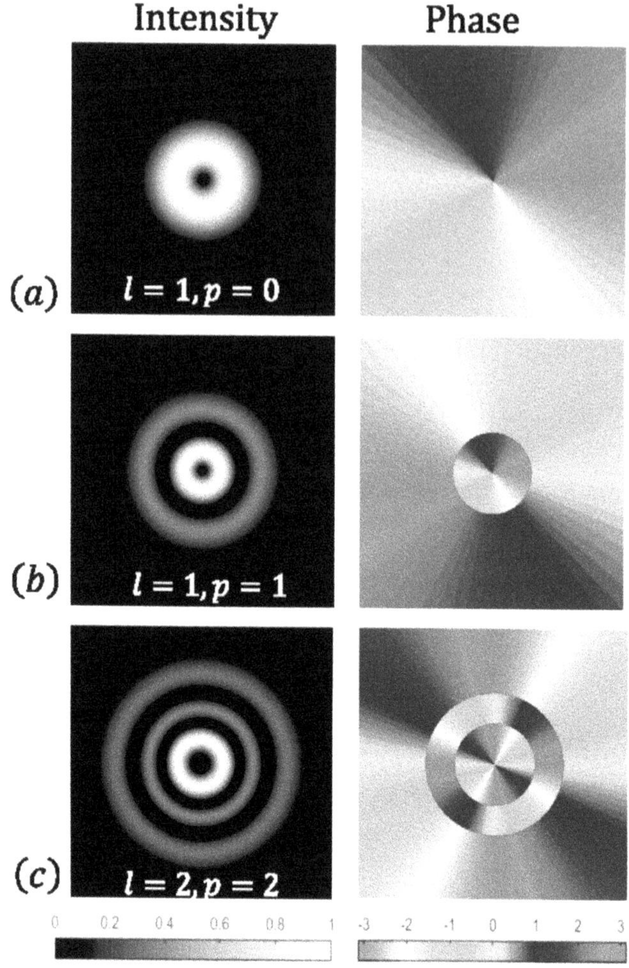



**Figure 5.1.** The intensity and phase of different Laguerre–Gauss modes in the $z = 0$ plane is shown. The azimuthal ($l$) and radial ($p$) indices take the values: (a) $l = 1, p = 0$, (b) $l = 1, p = 1$ and (c) $l = 2, p = 2$ respectively.

## 5.2 Orbital angular momentum of LG modes

Light carries both linear momentum and angular momentum. The linear momentum is equal to $h/\lambda$ per photon while the angular momentum depends on the beam's spatial structure and polarization. The linear momentum density $\boldsymbol{p}$ and the angular momentum density $\boldsymbol{j}$ of a light beam can be obtained from its electric and magnetic field as,

$$\boldsymbol{p} = \epsilon_o(\boldsymbol{E} \times \boldsymbol{B}), \tag{5.9}$$

$$\boldsymbol{j} = \epsilon_o[\boldsymbol{r} \times (\boldsymbol{E} \times \boldsymbol{B})] = (\boldsymbol{r} \times \boldsymbol{p}). \tag{5.10}$$

In the Lorentz gauge, the field $U(x, y, z)$ in equation (5.1) may be associated with a scalar component of the vector potential $A$. Substituting this field in the expressions for $p$ and $j$ above (equation (5.9) and (5.10)), one may obtain the linear momentum density of a light beam given by [6],

$$
\begin{aligned}
p &= \frac{\epsilon_o}{2}(E^* \times B + E \times B^*) \\
&= \frac{i\omega\epsilon_o}{2}[u^*\nabla u - u\nabla u^*] + \omega k \epsilon_o \,|U|^2\,\hat{z} + \frac{\omega\sigma\epsilon_o}{2}\frac{\partial|u|^2}{\partial r}\hat{\theta},
\end{aligned}
\tag{5.11}
$$

where $\sigma$ denotes the degree of polarization of light which is equal to $\pm 1$ for circularly polarized light and zero for linearly polarized light. From equations (5.10) and (5.11), we see that the $z$ component of the angular momentum density depends on the $\theta$ component of $p$ as $j_z = rp_\theta$. Therefore, in order to have a non-zero $j_z$ value, the light field must have a non-zero linear momentum in the azimuthal direction $(p_\theta \neq 0)$.

An infinite plane wave has only transverse components in its electromagnetic field and thus it cannot carry any angular momentum. However, beams of finite size can possess angular momentum in one of two ways. The angular momentum can contain a spin and an orbital contribution. The spin angular momentum (SAM) is associated with the photon spin and manifests itself in the form of circular polarization with a value of $\pm\hbar$ per photon [7, 8]. On the other hand, OAM is associated with the complex spatial profile of the light beam. As discussed in the previous section, optical vortex beams have an azimuthal phase structure given by $\exp(il\theta)$. This gives rise to a circulating component in the $k$-vector and correspondingly a non-zero OAM [9, 10]. As the OAM depends on the position vector $r$, vortex-free beams with non-zero net OAM can exist. For example, a Gaussian beam would have non-zero OAM if the axis around which OAM is being calculated is different from the beam's axis. However, these beams have un-quantized OAM. The angular momentum for circularly polarized LG modes in the $z$ direction is equal to $\sigma\hbar$ per photon for the SAM and $l\hbar$ per photon for the OAM [11, 12]. The value of SAM is independent of the choice of the $r$ and is therefore called intrinsic. It was shown by Berry [13] that in some cases, the OAM also does not depend on the choice of the beam axis and can be considered to be intrinsic.

## 5.3 Topological charge of OAM carrying beams

Consider an OAM carrying LG mode with $l \neq 0$ and $p \neq 0$. We observe that at $r = 0$ the field amplitude is zero and as a result the phase is not defined on-axis. Such a zero amplitude point is called a phase singularity if the integral of the phase gradient on a closed path $C$ enclosing the point is not equal to zero [14]:

$$
\oint_C \nabla\Theta \cdot dr \neq 0.
\tag{5.12}
$$

Here $\Theta$ is the phase of the LG mode defined in equation (5.5). In the neighborhood of the phase singular point, the gradient of phase, therefore, has a non-zero curl. For the LG modes, this phase change is an integer multiple of $2\pi$ and this integer is commonly known as the *topological charge* [11]. The topological charge denoted by $l$, is defined as [2]:

$$l = \frac{1}{2\pi} \oint_C \nabla\Theta \cdot dr. \tag{5.13}$$

The topological charge can take both positive and negative values. The sign of the topological charge describes the handedness of the helical phase. The phase increases in an anti-clockwise sense for a positive $l$ vortex. There can also be cases when $l$ is not an integer. Such a fractional charge beam can be expressed as a linear combination of the LG modes. The phase distribution of a LG(0, $l$) mode of charge $l$ in $z = 0$ plane is given by $\Theta(r, \theta, 0) = l\theta + k_z z = l\theta$ thus giving a phase gradient of $\nabla\Theta = \hat{\theta}/r + k_z\hat{z}$. We see that in any transverse plane the phase gradient has only an azimuthal component. This circulating phase gradient is shown in figure 5.2. The phase gradient $\nabla\Theta$ points in the direction of maximum increase of phase and is normal to the phase contour surface. Besides this, one can also look at the phase contour lines in figure 5.2. These equiphase lines are seen to terminate on the optical vortex in this paraxial propagation model. A peculiarity of the phase front of the OAM beam is the phase structure of form $l\theta$ which makes the propagating wavefront surface helical in nature. There is one major difference in the wavefront of these beams as compared to other non-singular beams. For other non-singular beams, the equiphase surfaces are separated in space from one another, whereas in singular beams there is no concept of distinct separate wavefronts, instead the helical wavefront is joined together to form a common spiraling surface stretching along the propagation direction. For optical vortices with $|l| > 1$, the wavefront structure has $|l|$ intertwined helical surfaces. The phase singularity is present only at a single

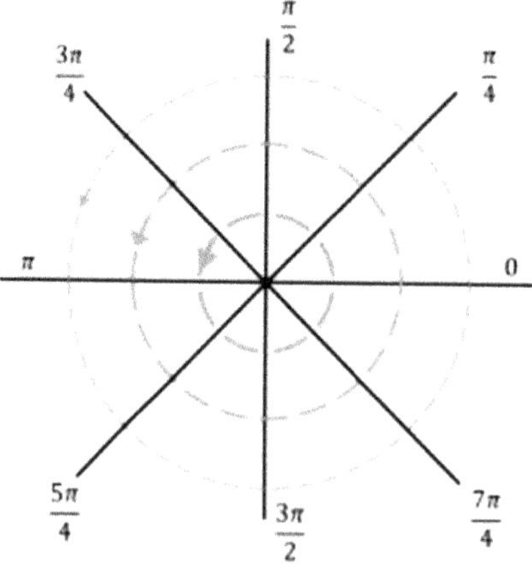

**Figure 5.2.** The transverse circulating phase gradient is shown (in orange) for $l = 1$ optical vortex. The strength of the phase gradient decreases as one moves radially away from the core. The phase contour lines representing constant phase are also shown (in black).

on-axis point and this manifests as lines of darkness on propagation in three-dimensional space.

## 5.4 Generation of OAM beams

Optical vortices occur naturally in speckles of coherent laser beams. These can also be observed in the dark regions of the three or more multiple beams interference pattern. However, in order to study the general structure and properties of optical vortices, it is first necessary to be able to generate pure OAM modes in a controlled manner in the laboratory. Only then, one can venture into harnessing their unique properties for specific applications. Over the last decades, many methods have been developed for generation of OAM beams. These methods can be broadly classified in two categories: (a) methods which try to manipulate the phase of an input beam to generate an azimuthal phase dependence and (b) modifying laser cavities so that output produced is itself a LG mode. There exists a large number of optical vortex generation methods, however, here we describe only those methods in detail which are relevant to the experiments reported here.

The first type of generation methods involve the use of specially designed phase elements, like spiral phase plates, wedges, diffractive elements like fork gratings, spiral zone plates, Dammann vortex gratings, adaptive helical mirrors and spatial light modulators (SLM) to modify the phase profile of the incoming beam to produce an azimuthal phase dependence. The LG modes generated from these methods are usually not 100% pure and might also contain unwanted higher order LG modes.

### 5.4.1 Spiral phase plate

Consider a case where a beam whose optical field is given by $u(r)$ is incident on a phase plate. For beams with small divergence and phase plates with sufficiently small height, the waves emerging from the plate remain in the paraxial regime. In this case, the operation of the phase plate can be considered to be an operation on the phase only and the transmitted field $u'(r)$ is given by:

$$u'(r) = u(r)\exp(i\psi), \tag{5.14}$$

where $\psi$ is the phase gained on passing through the plate. Therefore, in order to have an azimuthal phase term, it is natural to use a phase plate with helical shape. Spiral phase plate [15–17] is such an optical element and is one of the simplest method of introducing an azimuthal phase dependence in the input beam. Spiral phase plate is a transparent plate whose thickness varies proportional to the azimuthal angle $\theta$ around a point in the middle of the plate. The thickness along any radial line remains constant. This type of plate for optical wavelengths was first demonstrated by Beijersbergen *et al* [16] in 1994. Figure 5.3 shows the typical structure of such a phase plate. Let the height difference between the thickest and the thinnest portion of the spiral phase plate be $h$, then the thickness $t$ of the plate producing a positive charge vortex, can be written as a function of the azimuthal phase angle $\theta$:

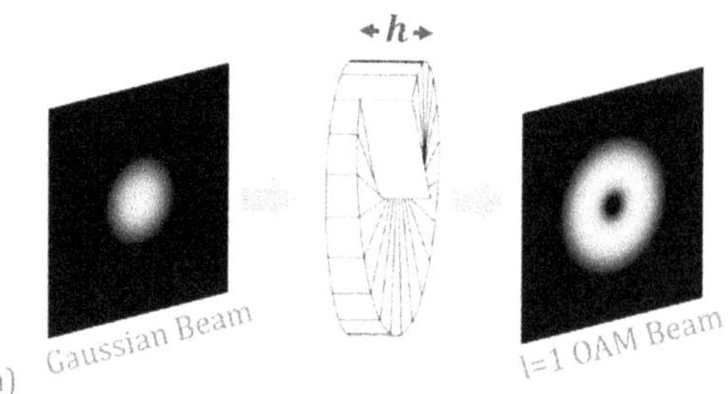


**Figure 5.3.** Spiral phase plate in action-conversion of Gaussian beam into $l = 1$ OAM state.

$$t(\theta) = \frac{h\theta}{2\pi}. \tag{5.15}$$

Assuming the vacuum wavelength of the input beam to be $\lambda$ and the refractive indices of the spiral phase plate and the background material to be $n$ and $n_o$ respectively, then the optical path length traversed by the beam is given by:

$$\text{Path Length} = \left(h - \frac{h\theta}{2\pi}\right)n_o + \frac{h\theta}{2\pi}n. \tag{5.16}$$

Neglecting the overall constant contribution, the phase $\psi$ gained by the beam on passing through the spiral plate is:

$$\psi(\theta) = k \times \text{Path Length} = \frac{h\theta}{\lambda}(n - n_o). \tag{5.17}$$

If the plate can be fashioned in such a way that the factor

$$\frac{h}{\lambda}(n - n_o) = l, \tag{5.18}$$

where $l$ is an integer, then, the field $u'(r)$ obtains a helical phase,

$$u'(r) = u(r)\exp(il\theta). \tag{5.19}$$

This beam on propagation acquires a central intensity null. Therefore, spiral phase plates can be used to convert non-helical beams into helical beams and can also be utilized to change the helicity of any given helical beam. However, the modes generated are not pure modes as only the phase of the incident beam is modified and not the amplitude distribution, for example, when an LG(0,0) beam falls on a $l = 1$ spiral phase plate, the azimuthal index ($l$) of the LG mode changes to $l = 1$, however it does not produce a pure LG(0,1) mode and in fact contains contributions from higher order LG modes of the same azimuthal order $l$ but different radial ($p$) orders. It has been shown that only 78.5% of the beam is in LG(0,1) mode [16]. The fabrication of a continuous phase ramp is very difficult at optical wavelengths and it is usually approximated as a staircase structure which causes further loss in mode purity. Another limiting factor is the dependence of the spiral phase plate on the input beam wavelength (see equation (5.18)). Recently new methods like photo-polymerization [18–20] and micro-machining techniques [21, 22] have been developed for manufacturing spiral phase plates.

## 5.4.2 Diffractive optics

This method involves the use of diffractive optical elements (DOE) to convert coherent non-singular beams to beams containing optical vortices. Two very well-known examples are fork grating [23, 24] and spiral zone plates [25, 26]. The fork design can be implemented as amplitude or phase gratings. In the case of fork grating, multiple orders ($\approx n$) with azimuthal dependence of $\exp(in\theta)$ are obtained with each diffraction order diffracted at a specific angle. This spatial separation of the orders makes it easier to spatially filter and use the desired vortex order. An example of fork grating and spiral zone plate for producing vortices of charges $l = 1$ and $l = 3$ is shown in figure 5.4. These DOE structures may be implemented in the form of computer-generated holograms (CGHs) [27–29]. The fork grating and the spiral zone plates can be generated by interfering a helical vortex beam with a plane and a spherical wave respectively [30]. The image of the obtained interference pattern can then be simply printed onto a film or mask, thus bypassing the need to physically record them. Just like spiral phase plates, the diffraction orders produced by these diffractive elements are not pure LG modes and contain superposition of many LG states [31]. The CGH transparency mask can, however, be used with multiple wavelengths.

**Figure 5.4.** The sinusoidal amplitude fork grating and spiral zone plate are shown for topological charges, (a) and (c) $l = 1$ and (b) and (d) $l = 3$. The fork grating is shown in (a) and (b) while (c) and (d) shows the spiral zone plate.

### 5.4.3 Spatial light modulators

The CGHs for generating vortex carrying beams can also be implemented on spatial light modulators (SLMs). SLM is a pixellated device that can modulate the phase, amplitude and polarization of the incident input beam [32–35]. There are various types of SLMs which differ from each other by the manner in which the phase modulation is controlled (electronically, optically or magnetically) [36–41]. The most common SLMs are the electronically controlled phase-only devices [42, 43]. The SLMs work in reflective mode and make use of liquid crystals to introduce the desired phase retardation into the input beam. The refractive index of the liquid crystals can be modified by applying a suitable electric field. The light reflecting from different pixels of the SLM device encounters different refractive indices, thus acquiring the desired phase difference between them. Many techniques have been developed to utilize these phase-only SLMs for carrying out amplitude [44, 45] as well as complex amplitude modulation [46, 47] of the input beam as required in several applications. Therefore, any arbitrary beam can be obtained by displaying the required CGH produced on the SLM. Unlike the traditional DOEs, SLMs offer a dynamic reconfigurable alternative for producing complex fields. Another interesting property of the liquid-crystal SLMs is that they are often sensitive to only one linear polarization. If the input beam is polarized at 45° to the preferred orientation (as may be specified by SLM manufacturer), then only the polarization component parallel to the SLM axis (assume $y$-polarization) interacts with the SLM and acquires the displayed phase pattern. The SLM acts like a plane mirror for the other ($x$-polarized) component of the beam and simply reflects it back without any phase modulation.

### 5.4.4 Mode converters

This method uses cylindrical lenses to convert a Hermite–Gaussian (HG) mode into a LG mode of the same order [48, 49]. A cylindrical lens changes the shape of the incoming wavefront in only one direction. Therefore, the beam would acquire two different Gouy phases along the $x$ and $y$ directions. By fine tuning this Gouy phase difference between a pair of HG modes, one can produce a pure LG mode. The second cylindrical lens is used to remove the asymmetry in the beam's curvature. Figure 5.5 shows the conversion of a diagonal HG mode into a LG mode.

### 5.4.5 Other methods

Optical vortices can also be generated by reflecting them from adaptive helical mirrors [50]. It is also known that three or more beam interference gives rise to dark points where complete destructive interference takes place. These points are actually optical vortices. Therefore, one can generate optical vortex arrays by multiple beam interference methods [51–53] by suitably fine tuning the tilt between the interfering beams. Direct generation of OAM modes from a laser cavity has also been demonstrated [54–58].

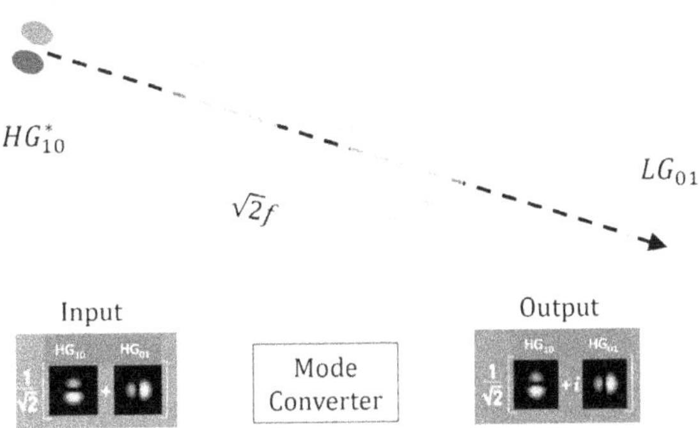


**Figure 5.5.** The conversion of a HG mode into a LG mode by a pair of cylindrical lenses is shown.

## 5.5 Detection of phase singularities

Detection of optical vortices is a vast and important research field in itself [1, 2] and some important detection schemes are provided in this section.

### 5.5.1 Interference based methods

Since the vortex beams are characterized with an azimuthal phase dependence, interferometry is a natural tool for detecting this phase profile. The vortex beam on interfering with a plane (spherical) beam produces interference fringes with a characteristic fork (spiral) pattern as shown in figure 5.4. The number of extra arms in the fork (spiral) pattern gives the value of the topological charge of the vortex beam. A Mach–Zender interferometer or lateral shear interferometer can be set up for testing a vortex beam.

### 5.5.2 Diffraction based methods

Phase also plays an important role in diffraction of beam through apertures. Vortex beams produce unique diffraction patterns on passing through apertures. These diffraction patterns are very distinct to the ones produced by non-singular beams and so can be used to identify a specific OAM state. Diffraction through different types of apertures like single slit [59], double slit [60], diffraction gratings [61], circular [62] and triangular apertures [63, 64], etc has been studied. The vortex beam diffracting through a triangular aperture produces a diffraction pattern with the number of bright lobes (on either side of the lattice) equal to the absolute value of $(l + 1)$ [63]. The helicity of the beam can also be identified by looking at the orientation of the diffraction pattern. Apertures of broken symmetry that is, apertures of shape other than square or circular produce diffraction patterns which give information about the charge of the beam. Recently, many such apertures, like elliptical annular [65], diamond shaped [66] and hexagonal apertures [67] have been studied.

### 5.5.3 OAM detection using lens aberrations

Lenses which suffer from Seidel aberrations like astigmatism, coma etc are useful for detecting vortex beams [2, 68, 69]. Lenses with circularly non-symmetric aberrations are more likely to break the symmetry of the vortex field and the resultant far-field intensity pattern can thus be used to identify the vortex beam and its charge.

### 5.5.4 Shack–Hartmann Wavefront Sensor

A Shack–Hartmann wavefront sensor is a device which consists of a two-dimensional array of lenslets of the same focal length [70]. Each of these lenslets produces a focal spot onto a two-dimensional detector array placed behind them. For an input plane beam, each of the focal spots lies in the center of each lens sub-aperture. However, this focal spot shifts from the center if any tilt or phase variation is present in the input beam. This collection of displacement of the individual focal spots is then used to estimate the wavefront of the input beam. This device has been successfully used as a vortex detector [71, 72]. It is important however to note that sufficient sampling of the wavefront by means of lenslet array is required.

## 5.6 Propagation dynamics of beams embedded with vortices

The propagation dynamics of optical vortices embedded into a host or carrier beam have been studied for linear, non-linear and turbulent media. Consider a host beam, for example a LG(0,0) mode, in which optical vortices have been introduced through one of the methods discussed earlier. These vortices can be centered on the host beam's axis or they can be distributed off-axis in the beam. In a two-dimensional transverse cross section of the beam, optical vortices appear as points of null intensity, but when propagated along the $z$-axis, these null intensity points form lines of darkness. The change in the position of the vortex inside the host beam in different $z$ planes is referred to as the propagation dynamics of the vortex. The trajectories taken by the optical vortices are greatly influenced by their placement in the beam, their neighboring optical vortices and the host beam's phase and intensity profile. Free-space propagation of an array of optical vortices nested in a Gaussian beam was numerically studied by Indebetouw [73] in the paraxial regime. This study showed that when the vortex array consisted of vortices of the same charge, the position of these vortices relative to each other and the host beam do not change on propagation. However, the pattern as a whole expands or contracts together with the host beam. The vortex pattern in the far field is also seen to undergo a rigid rotation by $\pi/2$ in the right (left) handed sense for the array of positively (negatively) charged vortices. In the case of two vortices of opposite charge, their trajectories are seen to depend on the relative value of their initial separation distance $2d$ and the host beam's waist $w_o$. Generally, the oppositely charged vortices tend to attract each other as they propagate from the waist and pairs of vortices starting sufficiently close to each other ($d < 0.5w_o$) may collide, interfere destructively and then all together vanish from the beam. Oppositely charged vortices which are sufficiently far away in the beam ($d > 0.5w_o$) might survive during propagation to the far field. A similar propagation study was reported in [74], where the dynamical behavior of optical vortices embedded in a plane wave was studied numerically using the angular spectrum method. It was observed that the perturbation caused in the plane wavefront of the beam due to the presence of a single vortex did not influence the propagation dynamics of this vortex. In this case, the vortex continued to propagate perpendicularly to the wavefront. In other words, what this means is that the vortices do not act on themselves and only the presence of nearby optical vortices alter their trajectories. In the same study, an isopolar (same charge) vortex pair was observed to gyrate around each other during propagation with their centroid propagating in the direction of the wave propagation. In the second case of a bipolar (opposite charge) vortex pair, the vortices were seen to drift laterally away from the direction of propagation. This pair did not gyrate and the separation distance between the vortices remained constant till the point where the vortices annihilated each other. The paper also mentions the sporadic creation and annihilation of bipolar vortex pairs near the focal points of converging wavefronts or when the interference occurs between different waves. Another study by Rozas *et al* [75] found contrasting differences between the propagation dynamics for $r-$ vortices and tanh vortices. The two types of vortices basically differ in the functional form of their core functions. The optical field for a charge $l$ vortex at $z = 0$ plane can be written as: $U(r, \theta, 0) = A(r, z = 0)\exp(il\theta)$ where $A(r, z = 0)$ describes the core or the amplitude function. The $r-$ vortices have amplitude which is linearly varying function of $r-$, for example

$$A(r, z = 0) = \left(\frac{r}{w_v}\right)^{|l|}, \tag{5.20}$$

where vortex core size is given by $w_v$. Similarly, the tanh vortices have localized core functions given by,

$$A(r, z = 0) = \tanh(r/w_v), \tag{5.21}$$

and are used in the description of an optical vortex soliton which has been observed in non-linear refractive media. If these two types of vortices are embedded into a Gaussian beam of some waist size $w_g$, then due to the arbitrary small core size of the tanh vortices, their peak intensity value is greater than that for the $r-$ vortices. Drawing on similarities between the hydrodynamic vortices and the optical vortices of paraxial wave fields, reference [75] described the important factors which influence the vortex trajectories. They observed that the transverse wave vector at the center of the optical vortex is unaffected by the vortex itself. The trajectory can only be affected by all other sources of phase and intensity gradient, such as diffracting waves and other vortices. This argument is similar to the theory put forward by Roux [74]. They further observed that like the rotation of point vortices in fluids, the tanh vortices when embedded into a Gaussian beam with their cores separated by a distance, orbit each other at rates which depend on the squared vortex separation distance. However, in a self-defocusing non-linear media, very high rotation rates could be achieved. A few other studies on vortex propagation dynamics have also been reported, for example, the cases of vortices with anisotropic phase profiles [76], and vortices in non-linear media [77, 78].

In this chapter we provided the basic description of the scalar OAM states of light and their properties such as topological charge, wavefront structure and propagation dynamics. The methods for generation and detection of OAM states were also discussed. When two distinct OAM states are embedded in orthogonal polarizations, the resultant beam is a vector beam containing a polarization singularity. In chapter 9 of the book, we will present a robust laser beam engineering principle using polarization singular beams. The concept of polarization singularities will be introduced in detail in the next chapter.

## 5.7 OAM modes as a communication basis

OAM modes as a spatial basis for multiplexing of channels has received much attention in recent years regarding optical and radio wave communication [79]. OAM allows angular momentum as a novel degree of freedom and as a result is a potential candidate to significantly increase information carrying capacity of communication systems [80–82]. We briefly examine this claim in this section.

In OAM based multiplexing of information, it is envisioned that OAM basis elements with common optic axis but different charge $l$ can be used to encode different channels. The transmitted wave-function can, for example, be a coherent or incoherent superposition of OAM modes and the information is embedded in time-varying coefficients representing the weights on the individual modes. The receiver may be preferably coaxial with the transmitter. We note that due to the relation:

$$\int_0^{2\pi} d\theta \ \exp[i(l_1 - l_2)\theta] = 0, \tag{5.22}$$

for integers $l_1 \neq l_2$, the OAM modes corresponding to the charge $l_1$ and $l_2$ are orthogonal to each other on any area centered on $r = 0$ and therefore can be considered at the input end as independent elements for the purpose of information transmission. This relation may suggest use of an arbitrarily large number of modes which may be used to encode information at the input end. As the different modes propagate to the receiver, we note that the energy in higher order modes spread over larger rings with on-axis nulls. It is well-known that the spot radius of the OAM mode with indices $(l, p)$ on propagation is given by [83]:

$$r_0(z) = W(z)\sqrt{l + 2p + 1}. \tag{5.23}$$

Here $W(z)$ is the spot size expressed in terms of beam waist $W_0$ as:

$$W(z) = W_0\sqrt{1 + \left(\frac{\lambda z}{\pi W_0^2}\right)^2}. \tag{5.24}$$

We therefore observe that the higher order modes denoted by $(l, p)$ have spot sizes that increase proportional to $\sqrt{l + 2p + 1}$. For simplicity let us assume that the on-axis detector radius $R_o$ is small compared to the beam waist $w_0$ at the detector. For the first two modes the fields are nominally defined as:

$$u_0 = A_0 \exp\left(-r^2/w_0^2\right), \tag{5.25}$$

and

$$u_1 = A_1 r \exp\left(-r^2/w_0^2\right)\exp(i\theta). \tag{5.26}$$

The power in the zero order mode collected by the detector given by:

$$P_0 = \int_0^{2\pi} \int_0^{R_o} d\theta \; r \; dr \; |u_0|^2, \tag{5.27}$$

is proportional to $\left[1 - \exp\left(-\frac{2R_o^2}{w_0^2}\right)\right]$. For the first order OAM mode the power $P_1$

on the same detector is however proportional to $\left[1 - \left(1 + \frac{2R_o^2}{w_0^2}\right)\exp\left(-\frac{2R_o^2}{w_0^2}\right)\right]$. In the

small detector ($R_o \ll w_0$) limit, the ratio $P_1/P_0$ is given by:

$$\frac{P_1}{P_0} = \left(\frac{R_o}{w_0}\right)^2. \tag{5.28}$$

The on-axis power received from first mode is therefore much smaller than the zeroth mode [84]. The possibility of detecting the higher order OAM modes on a finite sized receiver therefore depends on the signal-to-noise performance of the detection mechanism. From purely power considerations it is therefore clear that for a given signal-to-noise, the number of OAM modes that can be effectively detected by an on-axis receiver must be finite and cannot be made as large as we wish.

The question of information carrying capacity of any communication system boils down to examining the spatial degrees of freedom possessed by electromagnetic fields transmitted from and received by finite sized apertures [85]. For simplicity we assume that the OAM mode transmitter and receiver have apertures of radius $R_i$ and $R_o$ and are separated by a distance $z_0$ that is much larger than the transmitted and receiver apertures. The receiver, for example, may be considered in the Fraunhofer zone corresponding to the transmitter. Suppose a waveform $\psi_i(r, \theta)$ in the transmitter plane limited by the aperture of radius $R_i$. On propagation to the receiver plane in the Fraunhofer plane, the resulting wave-field is described as:

$$\psi_o(\rho, \phi) = C_0 \int\int_{R_i} r \; dr \; d\theta \; \psi_i(r, \theta) \exp\left[i2\pi r\frac{\rho}{\lambda z_0} \cos(\phi - \theta)\right]. \tag{5.29}$$

In the above equation $(r, \theta)$ and $(\rho, \phi)$ are polar coordinates in the transmitter and receiver planes respectively and $C_0 = i \exp(ikz_0)/(\lambda z_0)$. The two functions $\psi_i$ and $\psi_o$ are related by a 2D-Fourier transform due to the Fraunhofer approximation used here. Based on the earlier discussion of power received on the detector, we would like to have an input field such that it has maximal energy concentration within the receiver radius $R_o$. In other words we would like to maximize the energy concentration ratio:

$$E_r = \frac{\int\int_{R_o} \rho \, d\rho \, d\phi \, |\psi_o(\rho, \phi)|^2}{\int\int_{R_i} r \, dr \, d\theta \, |\psi_i(r, \theta)|^2}.$$

(5.30)

The optimal set of orthogonal functions that maximize this ratio are well-known to be the generalized prolate spheroidal functions (GPSF) [86]. An interesting property of GPSF is that they are eigenfunctions in the sense that if a GPSF is present in the transmit aperture, it will reproduce itself in the receive aperture with a constant multiplier $\mu$. Further, the number of significant eigenvalues $\mu$ is of the order of the space-bandwidth product of the system under consideration. The corresponding GPSF can retain maximal energy in the receiving aperture. In the finite transmitter and receiver geometry, OAM basis set is not necessarily the most optimal. In fact in terms of number of modes that may be transmitted, OAM basis may not have any particular advantage over other similar basis sets, for example, the Hermite–Gaussian basis set which is also a solution of the paraxial wave equation [87, 88].

At the receiver end the numerical aperture of the transmitting aperture is given by:

$$\text{NA} \approx \frac{R_o}{z},$$

(5.31)

which corresponds to the highest spatial frequency of

$$f_{\text{max}} = \frac{\text{NA}}{\lambda},$$

(5.32)

where $\lambda$ is the illumination wavelength. The basic space-bandwidth product considerations suggest to us that the two-dimensional degrees of freedom $N_0$ that can be supported between such a system are given by:

$$N_0 \approx \pi \frac{R_i^2 R_o^2}{\lambda^2 z^2}.$$

(5.33)

This would then be the maximum number of modes that may be supported by such a system irrespective of the basis set used for communication. The above considerations about degrees of freedom are for free space and in the presence of realistic turbulence even over several hundreds of meters of range, the OAM mode structure may get disturbed beyond recognition. The mode cross-talk in such cases will pose severe limitation for any communication system. We will not discuss this topic any further as it is not a major focus of the book. We will, however, discuss in detail the propagation of OAM states through atmospheric turbulence and present an interesting robust beam engineering principle later in the book that uses OAM states in orthogonal polarizations.

# References


[1] Gbur G 2015 *Singular Optics* (Boca Raton, FL: CRC Press)

[2] Senthilkumaran P 2018 *Singularities in Physics and Engineering Properties, Methods and Applications (IOP Series in Advances in Optics, Photonics and Optoelectronics)* (Bristol, UK: IOP Publishing)

[3] Nye J F, Berry M V and Frank F C 1974 Dislocations in wave trains *Proc. R. Soc. Lond. A Math. Phys. Sci.* **336** 165–90

[4] Goodman J W 2016 *Introduction to Fourier Optics* 3rd edn (Boston, MA: Roberts)

[5] Dennis M R, O'Holleran K and Padgett M J 2009 Optical vortices and polarization singularities *Prog. Opt.* **53** 293–363

[6] Allen L and Padgett M 2008 Introduction to phase-structured electromagnetic waves *Structured Light and Its Applications,* ed D L Andrews (Burlington: Academic) ch 1, pp 1–17

[7] Poynting J H 1909 The wave motion of a revolving shaft, and a suggestion as to the angular momentum in a beam of circularly polarised light *Proc. R. Soc. Lond. Ser.* A **82** 560–7

[8] Beth R A 1936 Mechanical detection and measurement of the angular momentum of light *Phys. Rev.* **50** 115–25

[9] Allen L, Padgett M J and Babiker M 1999 The orbital angular momentum of light *Progress in Optics* vol 39 (Amsterdam: Elsevier) ch 4, p 291

[10] Allen L, Beijersbergen M W, Spreeuw R J C and Woerdman J P 1992 Orbital angular momentum of light and the transformation of Laguerre-Gaussian laser modes *Phys. Rev.* A **45** 8185–9

[11] Soskin M S, Gorshkov V N, Vasnetsov M V, Malos J T and Heckenberg N R 1997 Topological charge and angular momentum of light beams carrying optical vortices *Phys. Rev.* A **56** 4064–75

[12] Courtial J, Dholakia K, Allen L and Padgett M J 1997 Gaussian beams with very high orbital angular momentum *Opt. Commun.* **144** 210–3

[13] Berry M V 1998 *Int. Conf. on Singular Optics* vol 3487, ed M S Soskin (Bellingham, WA: SPIE Optical Engineering Press) pp 6–11

[14] Soskin M S and Vasnetsov M V 2001 Singular optics *Progress in Optics* vol 42 (Amsterdam: Elsevier) ch 4, pp 219–74

[15] Kristensen M, Beijersbergen M W and Woerdman J P 1994 Angular momentum and spin-orbit coupling for microwave photons *Opt. Commun.* **104** 229–33

[16] Beijersbergen M W, Coerwinkel R P C, Kristensen M and Woerdman J P 1994 Helical-wavefront laser beams produced with a spiral phaseplate *Opt. Commun.* **112** 321–7

[17] Turnbull G A, Robertson D A, Smith G M, Allen L and Padgett M J 1996 The generation of free-space Laguerre-Gaussian modes at millimetre-wave frequencies by use of a spiral phaseplate *Opt. Commun.* **127** 183–8

[18] Knöner G, Parkin S, Nieminen T A, Loke V L Y, Heckenberg N R and Rubinsztein-Dunlop H 2007 Integrated optomechanical microelements *Opt. Express* **15** 5521–30

[19] Oemrawsingh S S R, van Houwelingen J A W, Eliel E R, Woerdman J P, Verstegen E J K and Kloosterboer J G 2004 Production and characterization of spiral phase plates for optical wavelengths *Appl. Opt.* **43** 688–94

[20] Sueda K, Miyaji G, Miyanaga N and Nakatsuka M 2004 Laguerre-Gaussian beam generated with a multilevel spiral phase plate for high intensity laser pulses *Opt. Express* **12** 3548–53

[21] Watanabe T, Fujii M, Watanabe Y, Toyama N and Iketaki Y 2004 Generation of a doughnut-shaped beam using a spiral phase plate *Rev. Sci. Instrum.* **75** 5131–5

[22] Yu Tsai H, Smith H I and Menon R 2007 Fabrication of spiral-phase diffractive elements using scanning-electron-beam lithography *J. Vac Sci. Technol.* B **25** 2068–71

[23] Basistiy I V, Bazhenov V Y, Soskin M S and Vasnetsov M V 1993 Optics of light beams with screw dislocations *Opt. Commun.* **103** 422–8

[24] Bazenhov V Y, Vasnetsov M V and Soskin M S 1990 Laser beams with screw dislocations in their wavefronts *JETP Lett.* **52** 429–31

[25] Sharma M K, Singh R K, Joseph J and Senthilkumaran P 2013 Fourier spectrum analysis of spiral zone plates *Opt. Commun.* **304** 43–8

[26] Vickers J, Burch M, Vyas R and Singh S 2008 Phase and interference properties of optical vortex beams *J. Opt. Soc. Am.* A **25** 823–7

[27] Heckenberg N R, McDuff R, Smith C P, Rubinsztein-Dunlop H and Wegener M J 1992 Laser beams with phase singularities *Opt. Quantum Electron.* **24** S951–62

[28] Heckenberg N R, McDuff R, Smith C P and White A G 1992 Generation of optical phase singularities by computer-generated holograms *Opt. Lett.* **17** 221–3

[29] Carpentier A V, Michinel H, Salgueiro J R and Olivieri D 2008 Making optical vortices with computer-generated holograms *Am. J. Phys.* **76** 916–21

[30] Senthilkumaran P, Masajada J and Sato S 2012 Interferometry with vortices *Int. J. Opt.* **2012** 517591

[31] Sacks Z S, Rozas D and Swartzlander G A 1998 Holographic formation of optical-vortex filaments *J. Opt. Soc. Am.* B **15** 2226–34

[32] Fisher A D and Lee J N 1986 The current status of two-dimensional spatial light modulator technology *Proc. SPIE* 0634 https://doi.org/10.1117/12.964024

[33] Efron U 1994 *Spatial Light Modulator Technology: Materials, Devices, and Applications (Optical Science and Engineering)* (Boca Raton, FL: CRC Press)

[34] Saleh B E and Teich M C 1991 *Fundamentals of Photonics* (New York: Wiley)

[35] Fukushima S, Kurokawa T and Ohno M 1991 Real-time hologram construction and reconstruction using a high-resolution spatial light modulator *Appl. Phys. Lett.* **58** 787–9

[36] Igasaki Y, Li F, Yoshida N, Toyoda H, Inoue T, Mukohzaka N, Kobayashi Y and Hara T 1999 High efficiency electrically-addressable phase-only spatial light modulator *Opt. Rev.* **6** 339–44

[37] Moddel G, Johnson K M, Li W, Rice R A, Pagano-Stauffer L A and Handschy M A 1989 High-speed binary optically addressed spatial light modulator *Appl. Phys. Lett.* **55** 537–9

[38] Clark R L 1997 Optically addressed spatial light modulator and method *US Patent* 5691836

[39] Hudson T D and Gregory D A 1991 Optically-addressed spatial light modulators *Opt. Laser Technol.* **23** 297–302

[40] Chung K H, Heo J, Takahashi K, Mito S, Takagi H, Kim J, Lim P B and Inoue M 2008 Characteristics of magneto-photonic crystals based magneto-optic SLMS for spatial light phase modulators *J. Magn. Soc. Jpn.* **32** 114–6

[41] Davis J A, Carcole E and Cottrell D M 1996 Intensity and phase measurements of non-diffracting beams generated with a magneto-optic spatial light modulator *Appl. Opt.* **35** 593–8

[42] Collings N, Davey T, Christmas J, Chu D and Crossland B 2011 The applications and technology of phase-only liquid crystal on silicon devices *J. Display Technol.* **7** 112–9

[43] Vaupotic N, Pavlin J and Cepic M 2013 Liquid crystals: a new topic in physics for undergraduates *Eur. J. Phys.* **34** 745

[44] Davis J A, Cottrell D M, Campos J, Yzuel M J and Moreno I 1999 Encoding amplitude information onto phase-only filters *Appl. Opt.* **38** 5004–13

[45] Márquez A, Iemmi C, Escalera J C, Campos J, Ledesma S, Davis J A and Yzuel M J 2001 Amplitude apodizers encoded onto Fresnel lenses implemented on a phase-only spatial light modulator *Appl. Opt.* **40** 2316–22

[46] Arrizón V, Ruiz U, Carrada R and González L A 2007 Pixelated phase computer holograms for the accurate encoding of scalar complex fields *J. Opt. Soc. Am.* A **24** 3500–7

[47] de Bougrenet de la Tocnaye J L and Dupont L 1997 Complex amplitude modulation by use of liquid-crystal spatial light modulators *Appl. Opt.* **36** 1730–41

[48] Beijersbergen M W, Allen L and van der Veen H E L 1993 Astigmatic laser mode converters and transfer of orbital angular momentum *Opt. Commun.* **96** 123–32

[49] Padgett M, Arlt J, Simpson N and Allen L 1996 An experiment to observe the intensity and phase structure of Laguerre–Gaussian laser modes *Am. J. Phys.* **64** 77–82

[50] Ghai D P, Senthilkumaran P and Sirohi R S 2008 Adaptive helical mirror for generation of optical phase singularity *Appl. Opt.* **47** 1378–83

[51] Masajada J and Dubik B 2001 Optical vortex generation by three plane wave interference *Opt. Commun.* **198** 21–7

[52] Vyas S and Senthilkumaran P 2007 Interferometric optical vortex array generator *Appl. Opt.* **46** 2893–8

[53] Boguslawski M, Rose P and Denz C 2011 Increasing the structural variety of discrete nondiffracting wave fields *Phys. Rev.* A **84** 013832

[54] Tamm C 1988 Frequency locking of two transverse optical modes of a laser *Phys. Rev.* A **38** 5960–3

[55] Tamm C and Weiss C O 1990 Bistability and optical switching of spatial patterns in a laser *J. Opt. Soc. Am.* B **7** 1034–8

[56] Rigrod W W 1963 Isolation of axi-symmetrical optical resonator modes *Appl. Phys. Lett.* **2** 51–3

[57] Ito A, Kozawa Y and Sato S 2010 Generation of hollow scalar and vector beams using a spot-defect mirror *J. Opt. Soc. Am.* A **27** 2072–7

[58] Kano K, Kozawa Y and Sato S 2012 Generation of a purely single transverse mode vortex beam from a he-ne laser cavity with a spot-defect mirror *Int. J. Opt.* **2012** 359141

[59] Ghai D P, Senthilkumaran P and Sirohi R S 2009 Single-slit diffraction of an optical beam with phase singularity *Opt. Lasers Eng.* **47** 123–6

[60] Sztul H I and Alfano R R 2006 Double-slit interference with laguerre-gaussian beams *Opt. Lett.* **31** 999–1001

[61] Moreno I, Davis J A, Melvin B, Pascoguin L, Mitry M J and Cottrell D M 2009 Vortex sensing diffraction gratings *Opt. Lett.* **34** 2927–9

[62] Ambuj A, Vyas R and Singh S 2014 Diffraction of orbital angular momentum carrying optical beams by a circular aperture *Opt. Lett.* **39** 5475–8

[63] Hickmann J M, Fonseca E J S, Soares W C and Chávez-Cerda S 2010 Unveiling a truncated optical lattice associated with a triangular aperture using light's orbital angular momentum *Phys. Rev. Lett.* **105** 053904

[64] Liu Y, Tao H, Pu J and Lü B 2011 Detecting the topological charge of vortex beams using an annular triangle aperture *Opt. Laser Technol.* **43** 1233–6

[65] Tao H, Liu Y, Chen Z and Pu J 2012 Measuring the topological charge of vortex beams by using an annular ellipse aperture *Appl. Phys.* B **106** 927–32

[66] Liu Y, Sun S, Pu J and Lü B 2013 Propagation of an optical vortex beam through a diamond-shaped aperture *Opt. Laser Technol.* **45** 473–9

[67] Liu Y and Pu J 2011 Measuring the orbital angular momentum of elliptical vortex beams by using a slit hexagon aperture *Opt. Commun.* **284** 2424–9

[68] Singh R K, Senthilkumaran P and Singh K 2007 The effect of astigmatism on the diffraction of a vortex carrying beam with a Gaussian background *J. Opt. A: Pure Appl. Opt.* **9** 543–54

[69] Kotlyar V V, Kovalev A A and Porfirev A P 2017 Astigmatic transforms of an optical vortex for measurement of its topological charge *Appl. Opt.* **56** 4095–104

[70] Platt B C and Shack R 2001 History and principles of Shack-Hartmann wavefront sensing *J. Refract. Surg.* **17** S573–7

[71] Chen M, Roux F S and Olivier J C 2007 Detection of phase singularities with a Shack-Hartmann wavefront sensor *J. Opt. Soc. Am.* A **24** 1994–2002

[72] Murphy K, Burke D, Devaney N and Dainty J C 2010 Experimental detection of optical vortices with a Shack-Hartmann wavefront sensor *Opt. Express* **18** 15448–60

[73] Indebetouw G 1993 Optical vortices and their propagation *J. Mod. Opt.* **40** 73–87

[74] Roux F S 1995 Dynamical behavior of optical vortices *J. Opt. Soc. Am.* B **12** 1215–21

[75] Rozas D, Law C T and Swartzlander G A 1997 Propagation dynamics of optical vortices *J. Opt. Soc. Am.* B **14** 3054–65

[76] Kim G-H, Lee H J, Kim J-U and Suk H 2003 Propagation dynamics of optical vortices with anisotropic phase profiles *J. Opt. Soc. Am.* B **20** 351–9

[77] Kivshar Y S and Luther-Davies B 1998 Dark optical solitons: physics and applications *Phys. Rep.* **298** 81–197

[78] Swartzlander G A 2001 *Optical Vortex Solitons (Springer Series in Optical Sciences* vol 82) (Berlin: Springer)

[79] Willner A E *et al* 2015 Optical communications using orbital angular momentum beams *Adv. Opt. Photon.* **7** 66–106

[80] Tamburini F, Mari E, Sponselli A, Thidé B, Bianchini A and Romanato F 2012 Encoding many channels on the same frequency through radio vorticity: first experimental test *New J. Phys.* **14** 033001

[81] Yan Y *et al* 2014 High-capacity millimetre-wave communications with orbital angular momentum multiplexing *Nat. Commun.* **5** 4876

[82] Edfors O and Johansson A J 2011 Is orbital angular momentum (OAM) based radio communication an unexploited area? *IEEE Trans. Antennas Propag.* **60** 1126–31

[83] Phillips R L and Andrews L C 1983 Spot size and divergence for Laguerre Gaussian beams of any order *Appl. Opt.* **22** 643–4

[84] Andersson M, Berglind E and Björk G 2015 Orbital angular momentum modes do not increase the channel capacity in communication links *New J. Phys.* **17** 043040

[85] Chen M, Dholakia K and Mazilu M 2016 Is there an optimal basis to maximise optical information transfer? *Sci. Rep.* **6** 22821

[86] Slepian D 1964 Prolate spheroidal wave functions, Fourier analysis and uncertainty—IV: extensions to many dimensions; generalized prolate spheroidal functions *Bell Syst. Tech. J.* **43** 3009–57

[87] Zhao N, Li X, Li G and Kahn J M 2015 Capacity limits of spatially multiplexed free-space communication *Nat. Photon.* **9** 822

[88] Gaffoglio R, Cagliero A, Vecchi G and Andriulli F P 2017 Vortex waves and channel capacity: hopes and reality *IEEE Access* **6** 19814–22

## Orbital Angular Momentum States of Light
Propagation through atmospheric turbulence
**Kedar Khare, Priyanka Lochab and Paramasivam Senthilkumaran**

# Chapter 6

# Introduction to polarization singularities

Polarization singularities form an important class of beam structures that can be generated using the orbital angular momentum (OAM) states. As discussed in the previous chapter the OAM in scalar light beams is generally attributed to phase singularities in the beam although beams with non-zero OAM that are non-singular also exist. When OAM states with different topological charge are embedded in orthogonal polarizations, the resultant beams have polarization singularities as we will discuss in detail in this chapter. In comparison to the subject of phase singularities, the complications arising due to additional parameters of polarization of light, makes the subject of polarization singularities relatively hard to follow and as a result the potential of polarization singularities has not been explored extensively so far. This chapter is aimed at providing an in-depth understanding of polarization singularities so that readers will appreciate this fascinating field. In the latter part of the book, we will introduce a robust laser beam engineering principle using polarization singular beams. This chapter will, therefore, provide the essential background for this purpose.

## 6.1 Polarization state of light beams

Consider a coherent monochromatic plane wave traveling along the $z$ direction. For such a wave the electric field oscillations are restricted to the $xy$ plane. As the beam propagates, the changes occurring to the tip of the electric field vector describe the state of polarization (SOP) of the beam [1]. If the electric field oscillations are such that the tip of the electric field vector draws a circle, an ellipse or a line in a projected $xy$ plane perpendicular to the beam propagation direction, the beam is said to be circularly, elliptically or linearly polarized respectively. For paraxial beams, we are mainly concerned with transverse fields that are described by two orthogonal polarizations in the $xy$ plane. The amplitude and phase of the component oscillations of a given state of polarization during orthogonal decomposition are crucial for determining the resultant polarization state. An equivalent description of

    6-1    

the state of polarization is in terms of Stokes parameters that do not deal with the component amplitudes and phases, but instead use directly measurable quantities as we describe next.

### 6.1.1 Stokes parameters

The state of polarization of light can be described using Stokes parameters [2, 3] that are expressed using intensity measurements as follows.

$$
\begin{aligned}
S_0 &= I_x + I_y, \\
S_1 &= I_x - I_y, \\
S_2 &= I_{45°} - I_{-45°}, \\
S_3 &= I_{RCP} - I_{LCP},
\end{aligned}
\tag{6.1}
$$

where $I_x$, $I_y$, $I_{(+45°)}$, $I_{(-45°)}$, $I_{(LCP)}$ and $I_{(RCP)}$ are the component intensities when a given SOP is decomposed into linear states oriented along $x$, $y$, $(+45°)$, $(-45°)$ and circular states—left circular polarization (LCP) and right circular polarization (RCP) respectively. Stokes parameters are related as:

$$
S_0^2 \geqslant S_1^2 + S_2^2 + S_3^2,
\tag{6.2}
$$

with the equality holding for fully polarized light. In this chapter we will describe polarization singularities in fully polarized light. For a linearly polarized light the Stokes parameter $S_3 = 0$ and $S_1^2 + S_2^2 = S_0^2$ ; for circularly polarized light both $S_1 = 0$ and $S_2 = 0$ and $S_3 = \pm S_0$. The polarization state is considered inhomogeneous if the Stokes parameters are functions of the transverse $(x,y)$ position coordinates. As we will discuss later, such an inhomogeneous polarization state occurs when the two orthogonal (and temporally coherent) polarization components of a light beam have a distinct spatially varying amplitude-phase structure.

### 6.1.2 Azimuth and ellipticity

Apart from Stokes parameters the state of polarization can be described by two other parameters—ellipticity $\chi$ and azimuth $\gamma$. For a fully polarized light ellipticity $\chi$ is defined by the relation

$$
\tan \chi = \pm \frac{b}{a},
\tag{6.3}
$$

where $a$ and $b$ are the major and minor axis of the polarization ellipse, the positive sign is for the right-handed ellipse and the negative sign is for the left-handed ellipse. The quantity $\chi$ varies between $-\pi/4$ and $+\pi/4$. The azimuth $\gamma$ is the orientation angle of the major axis of the ellipse with respect to the reference direction (usually $x$). The azimuth varies between 0 and $\pi$. In figure 6.1, various polarization ellipses with different azimuth or/and ellipticity are shown for illustration.

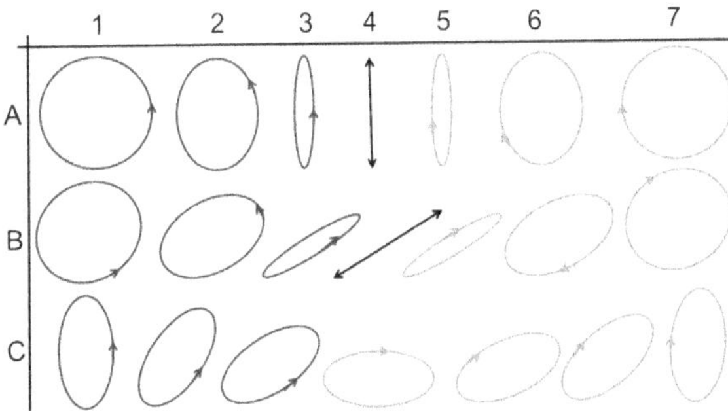

**Figure 6.1.** Illustration of polarization states with different azimuth and ellipticity. In row A, SOPs starting from right circularly polarized to left circularly polarized are shown corresponding to ellipticity of $+\frac{\pi}{4}$ to $-\frac{\pi}{4}$. Row B shows states having the same azimuth but varying ellipticity. All the figures in row C show states with same ellipticity but different azimuth.

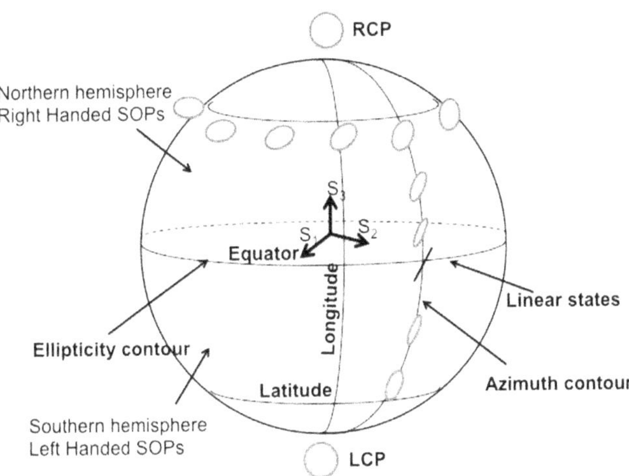


**Figure 6.2.** Poincaré sphere. North and south poles represent right and left circularly polarized light. Equatorial points represent linear polarization states and the rest of the points on the Poincaré sphere represent elliptically polarized light. The longitudes are azimuth contours and latitudes are ellipticity contours.

### 6.1.3 Poincaré sphere

Using the Stokes parameters, and from relation equation (6.2), it is natural to construct a sphere of radius $S_0$ in which $S_1$, $S_2$, $S_3$ form three orthogonal axes. This sphere is called the Poincaré sphere and every point on the surface of the sphere represents a particular state of fully polarized light (figure 6.2). Normalized Stokes parameters can be obtained by dividing each of the Stokes parameters by $S_0$. In the normalized coordinates the Poincaré sphere has unit radius. The north and south poles of the Poincaré sphere represent right and left circularly polarized light, equatorial points represent linearly polarized light with different azimuths and the remaining points represent elliptically polarized light. Points in the northern hemisphere are right-handed polarization states while points in the southern hemisphere are left-handed. We can also use two variables, namely the longitude and latitude [4] to represent any point on the surface of the Poincaré sphere. They are related to Stokes parameters as longitude

$$2\gamma = \arctan\left(\frac{S_2}{S_1}\right), \tag{6.4}$$

and latitude

$$2\chi = \arcsin\left(\frac{S_3}{S_0}\right). \tag{6.5}$$

While homogeneously polarized light can be represented by a point on the Poincaré sphere, it may be noted that inhomogeneously polarized light is represented by a collection of points or regions on the Poincaré sphere.

## 6.2 Decomposition of a general state of polarization

Light in any SOP can be decomposed into two linearly polarized orthogonal states. In other words, any SOP can be represented as a superposition of two linearly polarized states. Consider the following superposition

$$\vec{u} = a_x \exp(i\delta_x)\,\hat{x} + a_y \exp(i\delta_y)\,\hat{y}. \tag{6.6}$$

Here $a_x$, $a_y$ are the component amplitudes of the simple harmonic oscillations (for a coherent monochromatic light) occurring in $xz$ and $yz$ planes and $\delta_x$, $\delta_y$ are the component phases respectively. The SOP is circular when $a_x = a_y$ and $\delta_y - \delta_x = (2n + 1)\pi/2$ for an integer $n$. The SOP is linear when $\delta_y - \delta_x = n\pi$ irrespective of the values of $a_x$ and $a_y$. The SOP is elliptical in general when linear and circular polarization conditions are not met. Decomposition of SOP in terms of linear basis is explained in most common text books on polarization [1]. It is clear that SOP may also be represented in terms of any other orthogonal basis such as the circular basis [5, 6]. In this case the two component oscillations are clockwise and counter-clockwise rotating circular polarized light oscillations with amplitudes $a_R$ and $a_L$ respectively. The superposition state here is given by

$$\vec{u} = a_R e^{i\delta_R}\hat{e}_R + a_L e^{i\delta_L}\hat{e}_L, \tag{6.7}$$

where $\delta_R$ and $\delta_L$ are the phases of the component oscillations. In the circular decomposition, the linear states occur when the component oscillations have same amplitude, i.e. $a_R = a_L$ and elliptical states occur when the component oscillations are such that $a_R \neq a_L$. The orientation of the plane of polarization or the orientation of the major axis of the ellipse is decided by the phase difference between two circular components, $\delta_R - \delta_L$. Table 6.1 describes the difference between the two types of decompositions as presented here. It is important to note that the component amplitude decides the azimuth in linear decomposition, whereas the component amplitudes decide the ellipticity in the case of the circular decomposition. Likewise the component phases control the ellipticity and the azimuth in the

**Table 6.1.** Comparison of decomposition using two different basis sets.

| S.No. | Decomposition | Component amplitudes | Component phases |
| --- | --- | --- | --- |
| 1 | Linear decomposition | Azimuth control | Ellipticity control |
| 2 | Circular decomposition | Ellipticity control | Azimuth control |

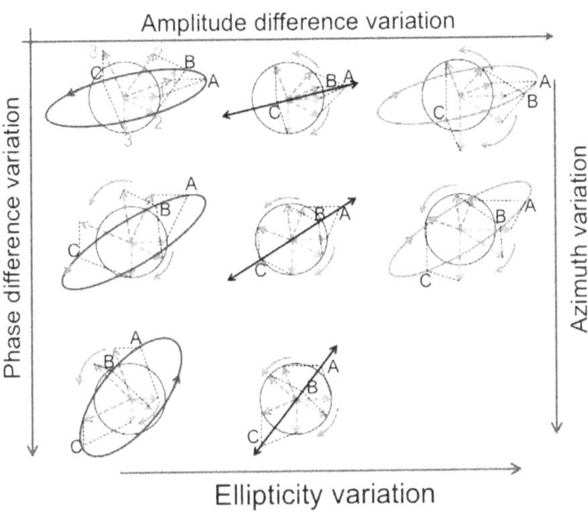


**Figure 6.3.** Azimuth control and ellipticity control by component phase and amplitude variation, respectively in circular basis.

two types decomposition. The case of circular polarization basis is depicted in figure 6.3 for illustration. The plots along each row in this figure are drawn with the azimuth kept constant. On the other hand, the plots along the column direction are drawn with the ellipticity kept constant. The component amplitudes shown by red color vectors rotate in a counter-clockwise direction while the component amplitudes in blue rotate in a clockwise direction. The sense of rotation of component vectors are shown by green circulating arrows. As time progresses, the orientations of the circular basis components change in anti-clockwise and clockwise directions for red and blue colored vectors respectively. At each time instance addition of vectors numbered as red 1 and blue 1 give the resultant vector denoted by A. Similarly vectors red 2 and blue 2 add to give resultant vector B at a later time. The vector C can be constructed in similar fashion with vectors numbered 3. This way the tip of the vector draws an anti-clockwise rotating ellipse and hence the resultant SOP is a right-handed elliptical state. In all the figures the resultant vectors rotate from A to C with increasing time. The component amplitudes are depicted by vectors of different lengths and phase difference between the component vectors are shown by different angular positions, the vectors (red 1 and blue 1) taken initially.

### 6.2.1 Helicity and spin

Helicity and spin are different [7, 8] and are not to be treated as synonymous to each other. Photons have integer spin (spin $\pm$ 1) with spin angular momentum (SAM) of $\hbar$ for a right circular polarized state and $-\hbar$ for a left circular polarized state. In the circular basis decomposition equation (6.7), we have seen that any polarization state of light is shown as a superposition of right and left circular polarization components (superposition of positive and negative spin states). The component amplitudes decide the helicity (or handedness) of the superposition state. If the right circular polarization component is larger than the left circular polarization component ($a_R > a_L$), then the resulting elliptical polarization state is said to right-handed ellipse state and so on. For linear polarization both the right and left circular polarization components are equal ($a_R = a_L$) and therefore there is no handedness associated with the linearly polarized states. If ($a_R = 0$), the light is left circularly polarized. Therefore, a left elliptically polarized light has negative helicity (handedness) but has both positive and negative spin components.

### 6.2.2 Homogeneous and inhomogeneous polarization distributions

Homogeneous polarization refers to uniform polarization distributions in the transverse cross-section of the beam. For example, a circularly polarized beam has homogeneous polarization distribution such that at all points in the beam the cross-section of the SOP is the same. Similarly a linearly polarized or an elliptically polarized light refers to uniform polarization across the beam. If a homogeneously polarized beam is passed through a polarizer, then there is a uniform reduction in the intensity of the beam, but there is no change in its spatial amplitude structure. The beams for which there is spatial variation of SOP across the beam cross-section are said to have inhomogeneous polarization and are also referred to as vector beams. Figure 6.4 shows the SOP distribution in a homogeneous ellipse field and inhomogeneous ellipse field for illustration. Beams with slowly varying polarization

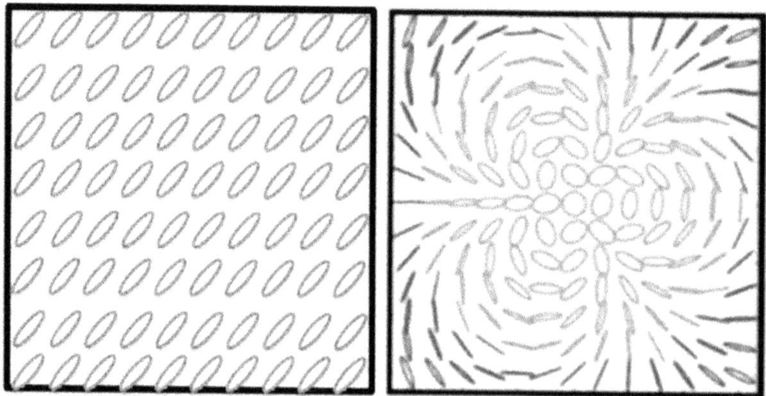

**Figure 6.4.** Homogeneously polarized beam (elliptically polarized light) and inhomogeneously polarized ellipse field.

distributions [9–11] have attracted interest in recent years. For such beams each of the parameters used to define the SOP is also smoothly varying. For example, the ellipticity can be spatially varying and the ellipticity distribution can have features such as extrema, saddles and so on. Since there are a number of parameters such as four Stokes parameters, ellipticity, azimuth, handedness or helicity associated with polarization of light, one can realize the occurrence of diverse topological features and possibilities in the study of inhomogeneous polarization. This makes the field more interesting and rich with an increased level of complexity. When the SOP varies point by point, it is normally illustrated by drawing polarization ellipses at equal spatial intervals (samples). If an inhomogeneously polarized beam is passed through a polarizer, then depending on the angle of the polarizer, one obtains a different spatial structure of the beam [12]. The effect of the rotation of the polarizer on homogeneous and inhomogeneous polarization distribution is illustrated in figure 6.5. We observe that while both the initial intensity profiles have a similar donut-like structure, the effect of the linear polarizer on the two states is completely different.

## 6.3 Singularities in optical fields

A point within an optical field is called as a singular point if a physical parameter associated with the field is not defined or not-well behaved (e.g. goes to infinity) [13]. In the neighborhood of this point the field gradients are typically large in magnitude. In electromagnetic fields, at a phase singularity the phase is not defined and at a polarization singularity the polarization azimuth is not defined. For an elliptically polarized light azimuth refers to the angle that the major axis of the ellipse makes with respect to a reference direction (say $x$). For a linearly polarized light azimuth refers to the angle of orientation of the vibration plane with respect to the reference orientation. A phase singularity is accompanied with a neighborhood in which all phase values ranging from 0 to $m2\pi$ are present for integer $m$. Similarly in a polarization singularity the neighborhood polarization distribution is such that the

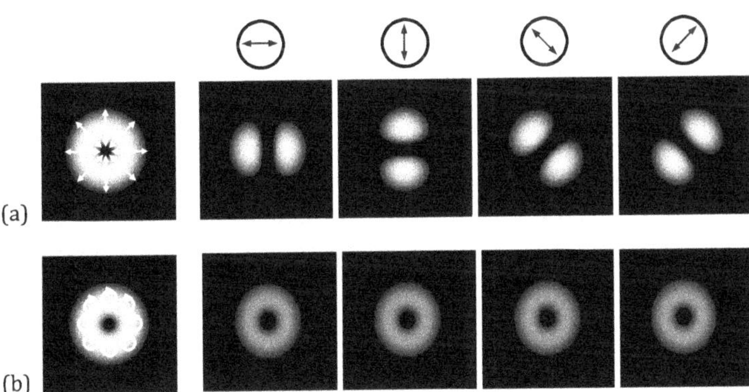

Figure 6.5. The effect of the rotation of the polarizer pass axis on the scalar and vector mode intensity is illustrated. Here (a) represents an input beam with radial polarization (inhomogeneous polarization) while (b) represents a right circularly polarized (homogeneous polarization) scalar beam.

polarization azimuth has all possible orientations. The phase gradient in a phase singularity and azimuth gradient in a polarization singularity circulate around the respective singularities. This also implies that in a phase singularity the phase contours converge at the singular point and in a polarization singularity the azimuth contours converge at the singular point.

### 6.3.1 Phase singularities

Optical phase singularity was already discussed in chapter 5 in the context of OAM states and is a point phase defect. The phase singularity is also referred to as optical vortex. The complex amplitude of a phase singular beam is nominally given by:

$$\tilde{U}(xf(r)\exp(i\phi)) = f(r)\exp(im\theta), \tag{6.8}$$

where $f(r)$ is the $r$-dependent part. The index $m$ is called the topological charge of the vortex defined by:

$$m = \frac{1}{2\pi} \oint \nabla\phi \cdot dl. \tag{6.9}$$

It can take positive and negative integer values depending on handedness of the helical wavefront. The phase distribution is given by azimuthally varying function $m\theta$, where $\theta$ is the angle in polar coordinate. The Laguerre–Gaussian (LG) beam with non-zero azimuthal index is an example of a phase singular beam. At $r = 0$ the amplitude of the LG beam is zero and thus the phase is undefined. Another commonly used example, is an $r$-vortex in which the complex amplitude is given by $\tilde{U} = (x \pm iy)^m$ where the amplitude distribution is given by $(\sqrt{x^2 + y^2})^m$ and the phase distribution is given by $\arctan(y/x)$. The $r$-vortex can be used to understand many of the properties of phase singularity. The transverse phase gradient [14, 15] for this vortex is given by $\nabla\theta = (m/r)\hat{\theta}$. The phase singular point is characterized by a circulating phase gradient and near the vortex core a large phase gradient exists [16–18]. As seen in the previous chapter, the phase singular beams carry OAM [19, 20]. This can be seen by analogy with quantum mechanics [21, 22] in which the momentum operator $L_z = -i\hbar\partial/\partial\theta$ acting on a wave function of the form $f(r)\exp\{im\theta\}$ gives rise to OAM of $m\hbar$. Here the charge $m$ of the vortex decides the angular momentum state. The phase distribution for a positive and negative unit charge vortex and the corresponding phase contours are shown in figure 6.6. The phase contours are radial from the vortex point. The phase gradient is normal to the phase contours and is therefore oriented in the azimuthal direction. The sense of circulation depends on which way the phase increases in the circulation. The amplitude profiles and distributions for a $r$-vortex, tanh-vortex and an LG beam are shown in figure 6.7.

### 6.3.2 Polarization singularities

The polarization structures of inhomogeneously polarized beams can be very complicated. Just like phase singularities of scalar fields, the inhomogeneously

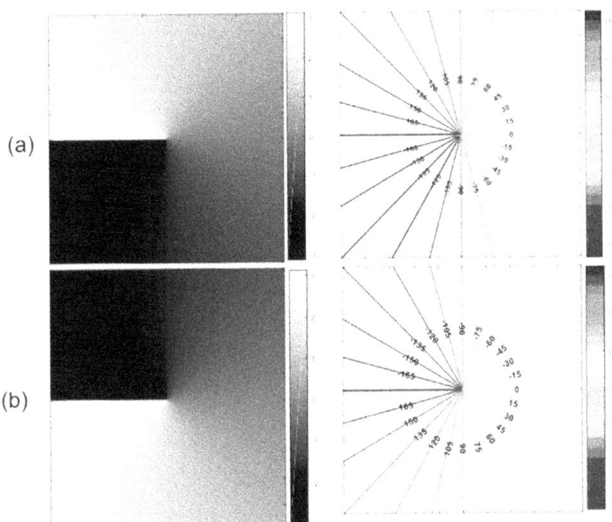

**Figure 6.6.** Phase distributions and phase contours for a vortex of charge (a) +1 and (b) −1. The numbers in the right panel figures are in degrees.

polarized field can also have its own singularities associated with polarization parameters like ellipticity or azimuth [9, 10, 12, 13, 23]. These are points in the vector field where some aspect of the beam's polarization is undefined. All vector beams which contain polarization singularities have spatially inhomogeneous polarization distribution. However, the reverse is not necessarily true, i.e. not all inhomogeneously polarized beams contain polarization singularities. The polarization singularities can be divided into two types—(a) elliptic point and (b) vector-point singularity depending on the state of polarization of the host beam. There also exits a third type of polarization singularity known as the Stokes point ($\Sigma$-point) singularity [24]. At a $\Sigma$-point, all three normalized Stokes parameters become undefined. Hence, at these points, the state of polarization itself is undefined. As of now very little is known about $\Sigma$-point singularities. In this chapter, we will only concentrate on the elliptic point and vector-point polarization singularities. Elliptic point singularities are present in inhomogeneous elliptically polarized fields [9, 10]. At these points, some property of the polarization ellipse becomes undefined. The planar elliptic point singularities are further characterized into two major categories —the L-lines and the C-points. The L-line is a continuous line of linear polarization embedded in a field of ellipses. The handedness of the ellipse (right or left) becomes undefined at the L-line. Hence, the L-line usually separates the regions of different handedness in an paraxial elliptically polarized field. The C-points are the points of circular polarization at which the orientation of the major (or minor) axis of the polarization ellipse is undefined. The polarization ellipses surrounding the C-point are usually arranged in three specific arrangements, known as the lemon, monstar and star. Different numbers of polarization lines terminate on these three structures. Only one line terminates on a lemon, three on a star and infinitely many with three

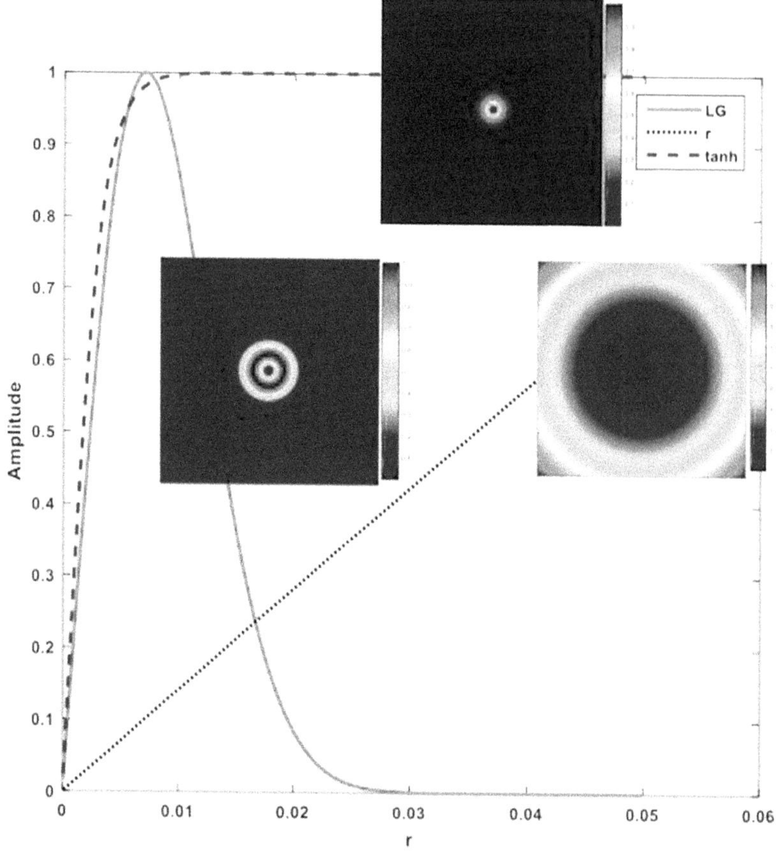

**Figure 6.7.** Three different types of amplitude profiles and the corresponding amplitude distributions for a unit charge vortex.

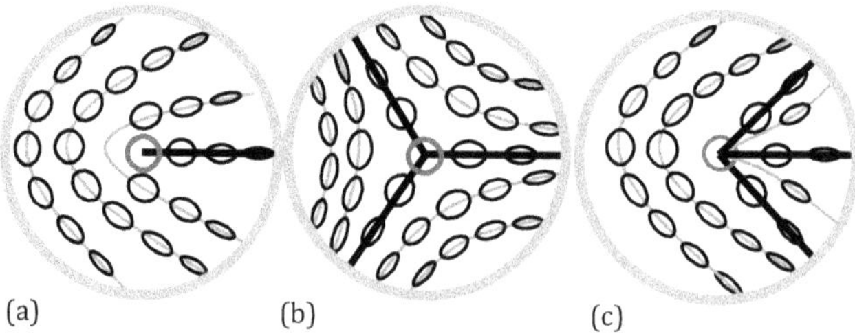



**Figure 6.8.** The morphology of C-type polarization singular structures. Here (a)–(c) show the lemon, star and monstar C-point singularities.

straight on a monstar [25]. The elliptical polarization fields for lemon, star and monstar are shown in figure 6.8. In order to explain the singularity index $I_c$ associated with the C-point singularities we show in figure 6.9 how the rotation of the major axis of the elliptical polarization changes around the singularity for

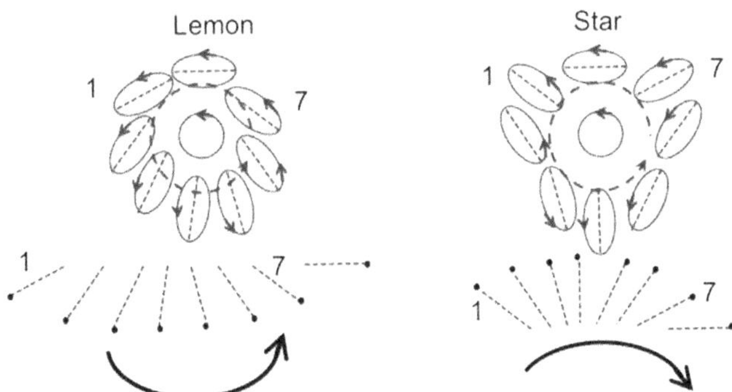



**Figure 6.9.** A lemon C-point and a star C-point. The sense of rotation of the major axis of the neighboring SOPs around the circular polarization state is opposite in the two cases. The amount of rotation is $\pm\pi$ radians.

positions 1–7. The sense in which the major axis of the ellipse rotates is seen to be opposite for the lemon and star type singularities. This information is captured by the index

$$I_c = \frac{1}{2\pi} \oint \vec{\nabla}\gamma \cdot d\vec{r}, \qquad (6.10)$$

which is a line integral of gradient of $\gamma$ (angle of orientation of the major axis of polarization ellipse) over a closed loop surrounding the C-point. As the $\gamma$ is defined modulo $\pi$, the $I_c$ is quantized in units of 1/2. Therefore, a C-point can be classified by its handedness (left-handed or right-handed) and index $I_c$. Lemon and monstar have $I_c = 1/2$ while star has $I_c = -1/2$. Monstar is a transitional singularity which only appears before creation or annihilation effects [23]. Monstar has the same index as a lemon and the same number of terminating lines as a star, that is why it is sometimes called a (le)monstar [25].

The vector-point singularities are isolated, stationary points of a linearly polarized vector field at which the orientation of the electric field vector is undefined. For a linearly polarized field, the handedness is undefined and for a V-point, even the orientation angle is undefined. Therefore, the field itself is zero at a V-point (note that this is not necessarily true for a C-point). The V-points are characterized using the Poincaré–Hopf index $\eta$ which is given by the line integral,

$$\eta = \frac{1}{2\pi} \oint \vec{\nabla}\gamma \cdot d\vec{r}, \qquad (6.11)$$

where the integral is evaluated over a closed loop surrounding the V-point. Azimuthally and radially polarized beams are lowest order V-point singular beams with $\eta = 1$ as shown in figures 6.10(a) and (c) respectively. The lowest order V-point singular beams with $\eta = -1$ are illustrated in figures 6.10(b) and (d).

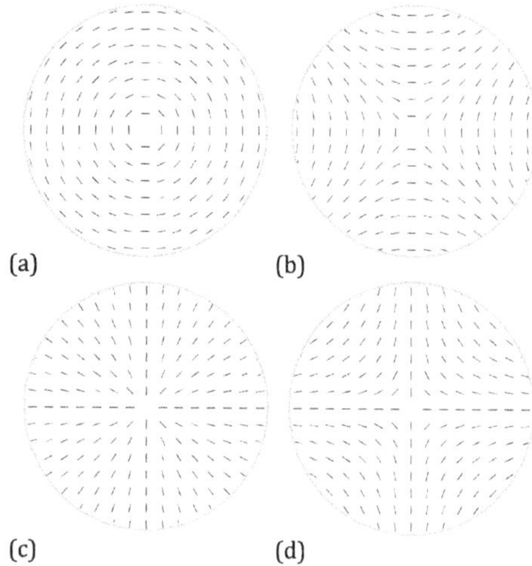



**Figure 6.10.** The morphology of V-type polarization singular structures. Here (a)–(d) shows type I, type II, type III and type IV V-point singularities. Type I and type III are commonly known as azimuthal and radial polarizations.

### 6.3.3 Polarization singularities as vector superposition of OAM states

So far polarization singularities have been explained in terms of the topological properties associated with the properties of polarization ellipses or orientations of linear polarization distributions. An explanation of this kind can sometimes leave an inexperienced reader somewhat confused as it is not clear how such states with exotic topological properties can be understood from a simpler picture of decomposing an arbitrary state as a superposition of two orthogonally polarized components. In circular basis representation of polarization states, the phase difference between orthogonal circular basis states decides the azimuth ($\gamma$) of the resulting polarization state. Hence to rotate the azimuth in the polarization distribution (like in figure 6.9), one has to have monotonically increasing or decreasing phase difference between circular basis components. Hence by having appropriate phase difference gradient in the component states, the required $\gamma$ variation can be realized in the polarization distribution. For the azimuth $\gamma$ to rotate in a clockwise direction the phase difference between the component beams should increase in a clockwise sense with respect to a point. One possibility to have such a phase difference is to superpose a negative charged vortex wave in RCP with a zero charge state in LCP. At a given radial distance from the vortex position, since the amplitude distribution in the two beams are not the same, during superposition at each point (on a given circle) the amplitude difference between the two beams leads to elliptical polarization. At the same time the phase difference between the two beams will lead to different orientations of the polarization ellipses. At the location of the vortex beam, the amplitude is zero while at the same location in the zero charge wave the amplitude is not zero. This leads to the circular polarization at the center in the superposition. The resulting polarization distribution is that of a C-point. Likewise, polarization singularities (C-points and

V-points) can be seen as superposition of two beams with different OAM in orthogonal circular polarizations [26].

Let us consider the orthogonal basis elements corresponding to the left and right circularly polarized $(\hat{e}_R, \hat{e}_L)$ light where $\hat{e}_R = (\hat{x} - i\hat{y})/\sqrt{2}$ and $\hat{e}_L = (\hat{x} + i\hat{y})/\sqrt{2}$. The general equation for a vector beam (on ignoring the $\exp(ikz)$ dependence) is then given by expression,

$$\vec{E}(r, \theta, z) = E_R(r, \theta, z)\hat{e}_R + E_L(r, \theta, z)\hat{e}_L. \tag{6.12}$$

In order to introduce a polarization singularity in the vector beam, the two transverse components $(E_R, E_L)$ are then taken to be the LG$(p, l)$ modes,

$$\vec{E}(x, y, z) = \frac{1}{\sqrt{2}}\left[\mathrm{LG}(p_1, l_1)e^{i\delta_1}\hat{e}_R + \mathrm{LG}(p_2, l_2)e^{i\delta_2}\hat{e}_L\right]. \tag{6.13}$$

As discussed earlier, the LG$(p, l)$ modes contain an on-axis phase singularity when $l \neq 0$. The C-point polarization singularities are generated when $l_1 \neq l_2$ whereas for V-point singularity the required condition is $p_1 = p_2$, $l_1 = -l_2$. The lowest order C-points $(l_1 = 1, l_2 = 0, \delta_1 = \delta_2 = 0)$ and V-points $(l_1 = 1, l_2 = -1, \delta_1 = 0)$ can be written for $p = 0$ as:

$$\vec{E}_C(x, y, z) = \frac{1}{\sqrt{2}}\left[\mathrm{LG}(0, 1)\hat{e}_R + \mathrm{LG}(0, 0)\hat{e}_L\right], \tag{6.14}$$

$$\vec{E}_V(x, y, z) = \frac{1}{\sqrt{2}}\left[\mathrm{LG}(0, 1)\hat{e}_R + \mathrm{LG}(0, -1)e^{i\delta_2}\hat{e}_L\right]. \tag{6.15}$$

The polarization singularities have rich topological structures with their own index conservation laws that follow from the behavior of the OAM modes in the orthogonal polarization states. A detailed discussion of these properties may be found in [13]. We observe that a beam carrying a C-point singularity has a net non-zero OAM while the total OAM of a V-point carrying beam is zero. The decomposition of polarization singular beams in circular basis is illustrated in figure 6.11.

In circular basis we see that each superposition of OAM states leads to a polarization singularity as given in table 6.2.

## 6.4 Stokes phase distribution and azimuth distribution

The polarization singularities can also be described using the concept of Stokes field. Using the Stokes parameters one may further define a mathematical construct called Stokes fields. For example using Stokes parameters $S_1$ and $S_2$ a complex field namely $S_{12} = S_1 + iS_2 = A_{12}\exp\{\phi_{12}\}$ can be constructed [9, 10, 27, 28]. The Stokes phase $\phi_{12} = \arctan(S_2/S_1)$ can be seen to be equal to $(\gamma/2)$. Hence, the phase vortices of the complex Stokes field $S_{12}$ are the polarization singularities. Therefore, constructing a Stokes field from the measured Stokes parameters (each parameter is a spatial distribution for an inhomogeneously polarized beam), is helpful in

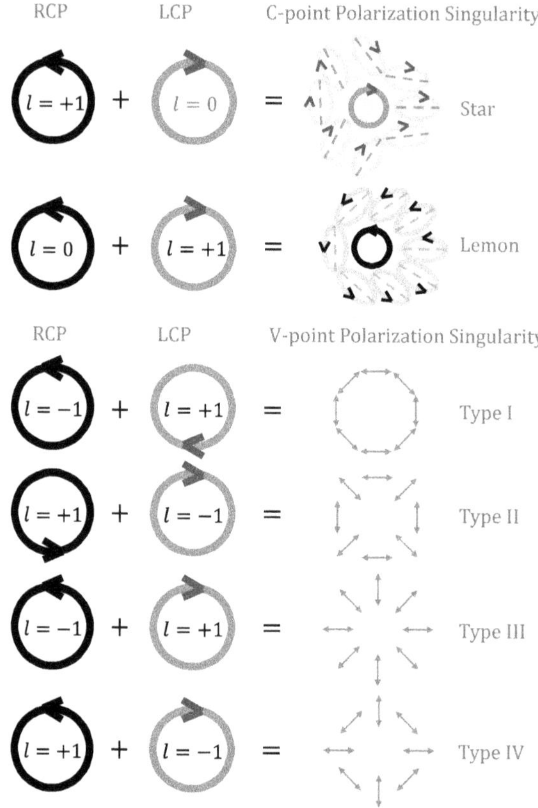


**Figure 6.11.** The figure illustrates the decomposition of C-point and V-point polarization singular structures in RCP–LCP ($\hat{e}_R - \hat{e}_L$) basis. Note that the right- and left-handedness is depicted by black and brown colors respectively. The C-point structures shown here are a left-handed star and a right-handed lemon. The green color depicts linearly polarized light.

identifying the polarization singularities. In figure 6.12(a) in the polarization distribution presence of two polarization singularities with opposite $I_C$ index can be identified as two vortex phase distributions in the Stokes phase. A V-point singularity and its Stokes phase distribution are shown figure 6.12(b). However the limitation in using Stokes fields is that the Stokes phase distribution does not distinguish the right- and left-handed C-point singularities and the dark or bright C-point singularities. Also this picture does not distinguish between integer charged C-points and V-point singularities. These have been illustrated in figure 6.12(c).

## 6.5 Generation and detection of polarization singularities

In the previous sections, the basics of polarization singularities have been explained. From the understanding of the decomposition of these singularities in orthogonal polarization bases, one can think of numerous techniques to realize them in the laboratory [4, 29–31]. By using a Mach–Zehnder, a Michelson, or a Sagnac configuration, the incident light is split into two orthogonal polarization states

**Table 6.2.** Superposition in circular basis. RCP-V means vortex in right circularly polarized light and LCP-V means vortex in left circular polarized light. $\Delta\phi$ is the constant phase shift given to one of the component beams. Note that in the superposition, in general when $l_1 \neq l_2$, dark C-point results as the amplitude variation in the two are different and the phase difference between them leads to rotating azimuth in the SOP distribution. Higher index C-points and V-points are also possible but are not tabulated here.

| RCP-V ($l_1$) | LCP-V ($l_2$) | $\Delta\phi$ | Polarization singularity | Type | Index |
|---|---|---|---|---|---|
| 1 | 0 | 0 | RH star | Bright C-point | $I_C = -\frac{1}{2}$ |
| 0 | 1 | 0 | RH lemon | Bright C-point | $I_C = +\frac{1}{2}$ |
| 0 | 1 | $\phi_0$ | RH lemon (rotated) | Bright C-point | $I_C = +\frac{1}{2}$ |
| -1 | 0 | 0 | LH lemon | Bright C-point | $I_C = +\frac{1}{2}$ |
| -1 | 0 | $\phi_0$ | LH lemon | Bright C-point | $I_C = +\frac{1}{2}$ |
| -1 | +1 | 0 | Radial polarization | V-point | $\eta = +1$ |
| -1 | +1 | $\pi$ | Azimuthal polarization | V-point | $\eta = +1$ |
| 1 | -1 | 0 | Type IV | V-point | $\eta = -1$ |
| 1 | -1 | $\pi$ | Type II | V-point | $\eta = -1$ |
| +1 | +2 | 0 | RH lemon | Dark C-point | $I_C = +\frac{1}{2}$ |
| 0 | 2 | 0 | RH (radial) ellipses | C-point | $I_C = +1$ |

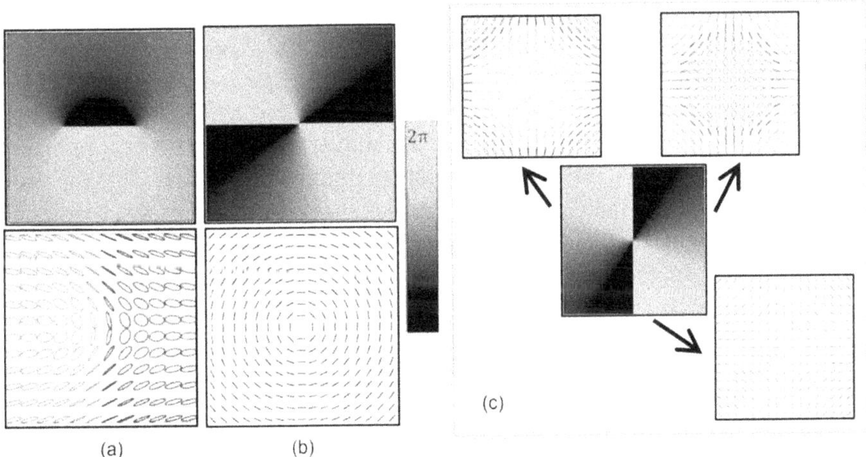

**Figure 6.12.** Stokes $\phi_{12}$ phase distributions and the corresponding SOP distributions. (a) A C-point dipole, (b) a V-point and (c) the Stokes phase which is same for two ellipse field singularities with $I_C = -1$ right-handed, left-handed and for a vector field singularity with $\eta = -1$.

and in either one or in both the beams, a phase vortex is inserted by using a vortex phase plate (or via an SLM). These two beams can then be combined back in circular basis to realize the polarization singularity. Another method is to use a single polarization element called an S-wave plate or q-plates [32, 33]. These plates have spatially varying polarization characteristics. Consider a half-wave plate that

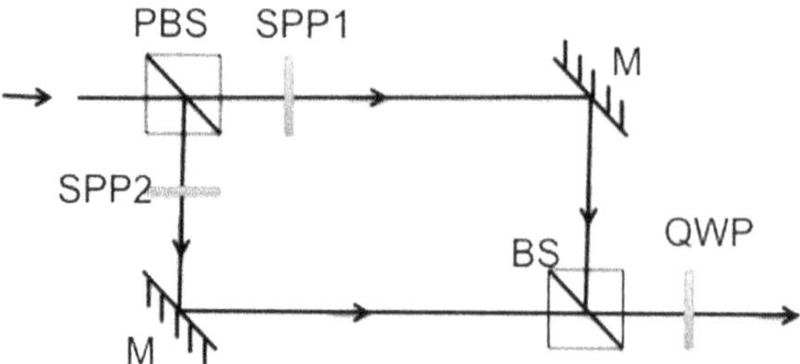


**Figure 6.13.** A schematic setup for the generation of C-point and V-point singularities.

can rotate the azimuth of the incident beam's SOP. By aligning the fast/slow axis of the half-wave plate to different directions at different spatial locations, the incident homogeneous light can be converted into a beam containing polarization singularity. Nano-structured materials [34] have also been designed to realize inhomogeneous polarization distributions that contain polarization singularities. Another interesting approach to generate polarization singularities uses stress engineered optics [35]. A schematic setup for the generation of polarization singularities is shown in figure 6.13. A 45° linearly polarized light is split equally at the polarizing beam splitter **BS**. Two spiral phase plates SPP1 and SPP2 inserted in the two arms of the interferometer implant a vortex in each of the beams, of charge say $l_1$ and $l_2$ in two beams respectively. The second beam splitter combines the two beams such that the two vortex cores exactly coincide, and combined beams propagate collinearly. A quarter wave plate QWP oriented at 45° ensures that the superposition is in circular polarization basis. When a $l_1 = -l_2$ V-point is generated; $l_1 = 0$ and $l_2 \neq 0$ or ($l_1 \neq 0$ and $l_2 = 0$) bright C-points are generated. When $l_1 \neq l_2$ and neither of them is of zero charge, a dark C-point is generated. There are many methods based on interference [36, 37] and diffraction [38, 39] for the detection phase singularities. But the number of methods available for the detection of polarization singularities are scarce at this point in time. One way to detect the polarization singularity is to obtain the spatially varying Stokes parameters using say a Stokes camera and analyzing the distributions or by drawing SOP distributions. There is one method, based on interferometry to detect the polarization singularities which is involved [40]. Recently diffraction and polarization transformation based methods [41] have been demonstrated for V-point detection.

## 6.6 Applications of polarization singular beams

Polarization singular beams have found some interesting applications. Radially polarized light has been shown experimentally and theoretically to focus to a spot which is significantly smaller than those achievable through linear or circular polarization states [42–45]. This effect is attributed to the strong longitudinal electric field component which is generated in the vicinity of the focus when a radially polarized light beam is passed through a high numerical aperture. In contrast, the azimuthally polarized light generates a purely transverse electric field which is zero at the center. Based on these unique properties of strong longitudinal electric field component and small focal spot size, radially polarized beams are being used in a variety of applications such as beam trapping, material processing (or laser based cutting) and optical memories. Polarization based spatial filtering can be done for directional and non-directional edge enhancement by making use of an s-wave plate [46]. The C-point polarization singular beams can be used to measure the optical activity in chiral samples [47] by detecting the singularity structure rotation that can be sensed by an analyzer. As we will discuss later in this book, beams containing C-point singularities are robust against intensity fluctuations on propagation through random medium in comparison with scalar Gaussian beams.

# References


[1] Hecht E and Zajac A 1974 *Optics* (Reading, MA: Addison-Wesley)

[2] Goldstein D H 2011 *Polarized Light* (Boca Raton, FL: CRC Press)

[3] Collett E 1993 *Polarized Light* (New York: Marcel Dekker)

[4] Ruchi, Bhargava Ram B S and Senthilkumaran P 2017 Hopping induced inversions and Pancharatnam excursions of C-points *Opt. Lett.* **42** 4159–62

[5] Maurer C, Jesacher A, Furhapter S, Bernet S and Marte M R 2007 Tailoring of arbitrary optical vector beams *New J. Phys.* **9** 78

[6] Pal S K, Ruchi and Senthilkumaran P 2017 Polarization singularity index sign inversion by half-wave plate *Appl. Opt.* **56** 6181–9

[7] Cameron R P, Barnett S M and Yao A M 2012 Optical helicity, optical spin and related quantities in electromagnetic theory *New J. Phys.* **14** 053050

[8] Cameron R P, Barnett S M and Yao A M 2014 Optical helicity of interfering waves *J. Mod. Opt.* **61** 25–31

[9] Dennis M R 2002 Polarization singularities in paraxial vector fields: morphology and statistics *Opt. Commun.* **213** 201–21

[10] Freund I 2002 Polarization singularity indices in Gaussian laser beams *Opt. Commun.* **201** 251–70

[11] Freund I, Soskin M S and Mokhun A I 2002 Elliptic critical points in paraxial optical fields *Opt. Commun.* **208** 223–53

[12] Rosales-Guzman C, Ndagano B and Forbes A 2018 A review of complex vector light fields and their applications *J. Opt.* **20** 123001

[13] Senthilkumaran P 2018 *Singularities in Physics and Engineering Properties, Methods and Applications (IOP Series in Advances in Optics, Photonics and Optoelectronics)* (Bristol, UK: IOP Publishing)

[14] Ghai D P, Senthilkumaran P and Sirohi R S 2008 Shearograms of optical phase singularity *Opt. Commun.* **281** 1315–22

[15] Roux F S 2003 Optical vortex density limitation *Opt. Commun.* **223** 31–7

[16] Nye J F and Berry M V 1974 Dislocations in wave trains *Proc. R. Soc. Lond. Ser. A* **336** 165–90

[17] Basisty I V, Soskin M S and Vasnetsov M V 1993 Optics of light beams with screw dislocations *Opt. Commun.* **103** 422–8

[18] Basisty I V, Soskin M S and Vasnetsov M V 1995 Optical wavefront dislocations and their properties *Opt. Commun.* **119** 604–12

[19] Allen L, Beijersbergen M W, Spreeuw R J and Woerdman J P 1992 Orbital angular momentum of light and the transformation of Laguerre-Gaussian laser modes *Phys. Rev. A* **45** 8185–9

[20] Allen L, Padgett M J and Babikar M 1999 *The Orbital Angular Momentum of Light (Progress in Optics* vol 39) (Amsterdam: Elsevier)

[21] van Enk S J and Nienhuis G 1992 Eigenfunction description of laser beams and orbital angular momentum of light *Opt. Commun.* **94** 147–58

[22] Nienhuis G and Allen L 1993 Paraxial wave optics and harmonic oscillators *Phys. Rev. A* **48** 656–65

[23] Gbur G 2015 *Singular Optics* (Boca Raton, FL: CRC Press)

[24] Freund I 2001 Polarization flowers *Opt. Commun.* **199** 47–63

[25] Dennis M R, O'Holleran K and Padgett M J 2009 Optical vortices and polarization singularities *Prog. Opt.* **53** 293–363

[26] Ruchi, Senthilkumaran P and Pal S K 2020 Phase singularities to polarization singularities *Int. J. Opt.* **2020** 2812803

[27] Ruchi, Pal S K and Senthilkumaran P 2017 C-point and V-point singularity lattice formation and index sign conversion methods *Opt. Commun.* **393** 156–68

[28] Pal S K and Senthilkumaran P 2019 Synthesis of stokes singularities *Opt. Lett.* **44** 130–3

[29] Tidwell W C, Ford D H and Kimura W D 1990 Generating radially polarized beams interferometrically *Appl. Opt.* **29** 2234–9

[30] Aadhi A, Vaity P, Chithrabhanu P, Redday S G, Prabhakar S and Singh R P 2016 Non-coaxial superposition of vector vortex beams *Appl. Opt.* **55** 1107–11

[31] Bhargava Ram B S, Sharma A and Senthilkumaran P 2017 Diffraction of V-point singularities through triangular apertures *Opt. Express* **25** 10270–5

[32] Machavariani G, Lumer Y, Moshe I, Meir A and Jackel S 2008 Spatially variable retardation plate for efficient generation of radially and azimuthally polarized beams *Opt. Commun.* **281** 732–8

[33] Marucci L, Manzo C and Paparo D 2006 Optical spin-to-orbital angular momentum conversion in inhomogeneous anisotropic media *Phys. Rev. Lett.* **96** 1639905

[34] Biener G, Niv A, Kleiner V and Hasman E 2002 Formation of helical beams by use of Pancharatnam–Berry phase optical elements *Opt. Lett.* **27** 1875–7

[35] Liang K, Ariyawansa A and Brown T G 2019 Polarization singularities in a stress-engineered optic *J. Opt. Soc. Am. A* **36** 312–9

[36] Vaughan J M V and Willetts D V 1983 Temporal and interference fringe analysis of excimer tem$_{01}$ laser *J. Opt. Soc. Am.* **73** 1018–21

[37] Ghai D P, Senthilkumaran P and Sirohi R S 2008 Detection of phase singularity using a lateral shear interferometer *Opt. Laser Eng.* **46** 419–23

[38] Moreno I, Davis J A, Melvin B, Pascoguin L, Mitry M J and Cottrell D M 2009 Vortex sensing diffraction gratings *Opt. Lett.* **34** 2927–9

[39] Ghai D P, Senthilkumaran P and Sirohi R S 2008 Single slit diffraction of an optical beam with phase singularity *Opt. Laser Eng.* **47** 123–6

[40] Angelsky O V, Mokhun I I, Mokhun A I and Soskin M S 2002 Interferometric methods in diagnostics of polarization singularities *Phys. Rev. E* **65** 1–5

[41] Bhargava Ram B S, Sharma A and Senthilkumaran P 2017 Probing the degenerate states of V-point singularities *Opt. Lett.* **42** 3570–3

[42] Dorn R, Quabis S and Leuchs G 2003 Sharper focus for a radially polarized light beam *Phys. Rev. Lett.* **91** 233901

[43] Quabis S, Dorn R, Eberler M, Glockl O and Leuchs G 2000 Focusing light to a tighter spot *Opt. Commun.* **179** 1–7

[44] Lerman G M and Levy U 2008 Effect of radial polarization and apodization on spot size under tight focusing conditions *Opt. Express* **16** 4567–81

[45] Zhan Q and Leger J R 2002 Focus shaping using cylindrical vector beams *Opt. Express* **10** 324–31

[46] Bhargava Ram B S, Senthilkumaran P and Sharma A 2017 Polarization-based spatial filtering for directional and nondirectional edge enhancement using an s-waveplate *Appl. Opt.* **56** 3171–8

[47] Samlan C T, Suna R R, Naik D N and Viswanathan N K 2018 Spin-orbit beams for optical chirality measurement *Appl. Phys. Lett.* **112** 031101

**IOP** Publishing

# Orbital Angular Momentum States of Light
Propagation through atmospheric turbulence
**Kedar Khare, Priyanka Lochab and Paramasivam Senthilkumaran**

# Chapter 7

# Theory of wave propagation in a turbulent medium

With the discussion of basic mathematical tools and description of OAM states of light as well as polarization singularities, we are now ready to tackle the problem of sending light beams through randomly fluctuating media like atmospheric turbulence. Long range beam propagation through turbulence is of interest to defense as well as free space optical communication (both classical and quantum) systems. While the study of laser beam propagation through the atmosphere has a long history over several decades, we believe that these studies have mainly focused on scalar beams. Secondly, there is a general lack of easily accessible but fairly rigorous beam propagation computation tools. The remaining part of this book will, therefore, first aim to explain the basic theory of light propagation through turbulence. Computational methodologies for propagation of arbitrary amplitude/phase/polarization structured beams will also be discussed in detail in coming chapters along with a easily accessible code base that may be used in standard packages like MATLAB or its open source clone Octave. Along the way we will also explain how to utilize the interesting diversity in the evolution of OAM states through turbulence that may then be utilized for engineering of robust laser beams.

## 7.1 Electromagnetic wave equation in random medium

The nature of propagation of monochromatic electromagnetic fields in free space has already been described in chapter 3. A turbulent medium like the atmosphere is characterized by spatially randomly varying dielectric constant $\varepsilon(\mathbf{r}, t)$. We assume that the atmosphere has constant magnetic permeability and zero conductivity implying an absence of free charges and external currents. We start with equation (3.19) which describes the propagation of the $E$-field in a medium with spatially varying refractive index.

doi:10.1088/978-0-7503-2280-5ch7          

$$\nabla^2 \boldsymbol{E} + \nabla(\boldsymbol{E} \cdot \nabla \ln \varepsilon) = \mu \frac{\partial^2}{\partial t^2}(\varepsilon \boldsymbol{E}). \qquad (7.1)$$

The second term on the left-hand side represents the coupling between the orthogonal polarizations of the electric field on propagation through random medium. We will calculate the magnitude of the depolarization term in the next section. For line of sight propagation, the magnitude of this term is very small at optical wavelengths and this term can be safely ignored when long range turbulence propagation is of interest. Let us assume that the time dependence of the electro-magnetic fields is sinusoidal of the form $\exp(-i\omega t)$ where $\omega$ is the frequency of the wave, so that the fields can be written as:

$$\boldsymbol{E}(r, t) = \boldsymbol{E}_o(r)\exp(-i\omega t), \qquad (7.2)$$

$$\boldsymbol{B}(r, t) = \boldsymbol{B}_o(r)\exp(-i\omega t). \qquad (7.3)$$

Usually the dielectric constant $\varepsilon(\mathbf{r}, t)$ is expressed as:

$$\varepsilon(r, t) = \langle \varepsilon(r) \rangle + \varepsilon_1(r, t), \qquad (7.4)$$

where the $\langle \varepsilon \rangle$ is the deterministic (non-random) average value of dielectric constant which can be a function of position. The random fluctuations about the mean value are contained in the term $\varepsilon_1$ which varies with both space and time. By definition, $\langle \varepsilon_1 \rangle = 0$ and $\langle |\varepsilon_1| \rangle \ll \langle \varepsilon \rangle$. This condition is valid, for example in the troposphere where $\langle \varepsilon \rangle \approx 1$ and $\langle |\varepsilon_1| \rangle \sim 10^{-6}\text{–}10^{-4}$. The dielectric constant can also be written in terms of the random refractive index function $n(r, t)$ as:

$$\varepsilon(r, t) = \varepsilon_o n^2(r, t). \qquad (7.5)$$

Where $n(r, t)$ is also a random function of space and time, defined in a similar fashion as equation (7.4):

$$n(\mathbf{r}, t) = \sqrt{\frac{\varepsilon(r, t)}{\varepsilon_o}} = \langle n(r) \rangle + n_1(\mathbf{r}, t), \qquad (7.6)$$

where $\langle n(r) \rangle$ gives the average value of the refractive index and $n_1(r, t)$ gives the fluctuations about the mean value. For small fluctuations, we get $\varepsilon_1(r, t) \approx 2n_1(r, t)$. For the moment, we will suppress the time dependence of the dielectric constant and will only pay attention to its spatial properties, therefore we assume $\varepsilon_1$ to be independent of time, i.e. $\varepsilon(r) = \langle \varepsilon(r) \rangle + \varepsilon_1(r)$. Such an assumption is valid as the atmospheric fluctuations are much slower compared to the electromagnetic frequencies. These fluctuations are induced by two distinct processes: (1) internal rearrangement of the turbulent eddies, and (2) due to the air stream which carries these eddies. If the velocity of the air stream which carries these eddies is very much greater than the turbulent velocity with which the eddies are mixing, then the structure of the turbulent air does not change and is simply carried forward along with the wind. This is known as the *Taylor's frozen flow hypothesis* and is usually invoked whenever temporal fluctuations of the atmosphere are to be studied. This allows us to take $\varepsilon$ out of the time derivative in equation (7.1) and we get:

$$\nabla^2 E_o(r) + \frac{\omega^2 n^2(r)}{c^2} E_o(r) = 0. \qquad (7.7)$$

This stochastic wave equation differs from the conventional wave equation through the fact that now the refractive index is a random function of $r$. Therefore, the equation is a homogeneous partial differential equation with space dependent coefficients. The refractive index function $n(r)$ varies from point to point in an unpredictable manner making it impractical to describe its value at all points in space and so it becomes imperative to describe the medium statistically. However, even for a given statistical description of $n(r)$, the exact solution for the wave equation is difficult to obtain and one has to apply approximate methods to get a practical solution. Just like in any other problem containing random variables, one can only attempt to solve equation (7.7) to obtain statistical average (or mean value) for the desired quantity of interest, for example, mean irradiance, average beam width, the variance of angle of arrival etc. The three components of the electric field vector in equation (7.7) can be solved independently from each other as we have already dropped the depolarization term which was responsible for the coupling between the three components. Therefore, we can decompose the equation (7.7) into three scalar equations for each component and ignore the vector nature of the electric field for the time being. Let $U(r)$ represent one of the scalar components transverse to the direction of propagation, then we can write equation (7.7) in the form commonly known as the *stochastic Helmholtz equation*:

$$\nabla^2 U(r) + k^2 n^2(r) U(r) = 0, \qquad (7.8)$$

where $k = \omega/c = 2\pi/\lambda$ is the wavenumber. Before we proceed to look into the various theoretical methods that have been employed to solve the above equation, we must first be able to provide the statistical description of the refractive index fluctuations denoted by $n_1(r)$. Early attempts to solve the stochastic Helmholtz equation were based on the geometrical optics method. In the geometrical optics method, diffraction effects are ignored and the phase fluctuations are calculated and, as per Snell's law a ray passes through regions of different refractive index variations. The curvature of the turbulent eddies causes focusing or de-focusing of rays as if they are passing through converging or diverging lenses which results in amplitude fluctuations. The results of this method are valid when the width of the scattering cone is smaller than the size of a turbulent eddy. This is usually stated in terms of the Fresnel length $\sqrt{\lambda L}$, (propagation distance $L$), which needs to be much smaller than the size of the turbulent eddies. Otherwise the diffraction effects can become important and the amplitude fluctuations are not correctly predicted [1]. Before providing details of models for light propagation through turbulence, it is important to understand the nature and physical origins of refractive index fluctuations in the atmosphere as we will discuss in the next section.

## 7.2 Description of the refractive index fluctuations in atmosphere

While it is common knowledge that the refractive index of atmosphere fluctuates with time, the nature of these fluctuations, their space–time scales and the magnitude of index fluctuations is important to understand before we can begin to describe sophisticated models of beam propagation in turbulence.

### 7.2.1 Origin of fluctuations in the index of refraction

The solar energy during daytime heats up the Earth's surface along with the surrounding air. It is well known that the air closer to the ground becomes warmer due to additional radiative heat transfer from the Earth's surface forming a vertical temperature gradient in the atmosphere. The hotter air is lighter or less dense than the cool air. This results into a rapid ascent of the hotter air into the atmosphere while the surrounding cooler air descends to take its place. This turbulent mixing of air in the atmosphere breaks down large air masses into randomly distributed pockets, each with its own characteristic temperature. The local index of refraction of air being sensitive to temperature, develops a random profile.

The turbulent mixing of the air gives rise to randomly distributed regions of high or low refractive index which are known as turbulent eddies. The eddies can be thought of as random lenses having different shapes and spatial sizes which are randomly moving in the atmosphere. However, the refractive index is also varying within an eddy as there are usually no refractive index discontinuities present in the atmosphere. The refractive index changes due to atmospheric turbulence are sufficiently small and the light beam passing through the turbulent eddies can be safely assumed to experience only phase changes with no accompanying amplitude changes. The phase fluctuations experienced by the input light beam are random and can give rise to focusing or de-focusing effects and beam wandering effects over large propagation distances. The phase fluctuations accumulated by the beam wavefront result in random destructive and constructive interference, that on propagation through multiple eddies give rise to fluctuations in the intensity profile of the beam, an effect commonly known as beam scintillation.

Kolmogorov's theory of atmospheric turbulence draws on similarities from fluid dynamics [2]. In the study of motion of a viscous fluid, two states can be observed; laminar and turbulent. Reynolds used a non-dimensional quantity known as the Reynolds number $\mathrm{Re} = v_{\mathrm{avg}} l_s / k_v$ to characterize these two types of fluid motion. Here, $v_{\mathrm{avg}}$ is the average velocity of the viscous fluid and $l_s$ denotes the characteristic dimension (or size) of the flow. The kinematic viscosity of the fluid is given by $k_v$. The fluid flow changes states from laminar to turbulent as the average velocity of the flow is increased. In laminar flow, the fluid follows a smooth path and the velocity characteristics are uniform or change in some deterministic fashion. However, when the Reynolds number exceeds a critical value, the flow becomes turbulent and dynamic mixing of the fluid occurs and leading to formation of turbulent eddies. For atmosphere, taking $v_{\mathrm{avg}} = 2\ \mathrm{ms^{-1}}$, $l_s = 10\ \mathrm{m}$ and $k_v = 1.5 \times 10^{-5}\ \mathrm{m^2\ s^{-1}}$, the Reynolds number is approximately given by $\mathrm{Re} = 1.3 \times 10^6$ which is very high (for laminar flow, the Reynolds number is typically up to order of $10^3$), and so the atmosphere shows turbulent air mixing.

In order to model optical turbulence, it is important to describe the statistics of the turbulent eddies. The result of the turbulent mixing of air is that the energy injected into large scale air masses is transferred to air masses on smaller scales. Richardson developed the *energy cascade theory* to visualize this energy transfer as shown in figure 7.1 [3, 4]. As the larger eddies break into small sized eddies, there is a continuum of eddy sizes. The eddies which are treated as homogeneous and isotropic are bounded by the eddy sizes $L_o$ and $l_o$. These sizes denote the outer and inner scale of the atmospheric turbulence. Eddies of sizes larger than $L_o$ are no longer isotropic and their structure is not well defined while eddies of sizes smaller than $l_o$ lose most of their energy heat generated through the viscous dissipation processes. As the eddies become smaller and smaller, the relative amount of energy injected and energy dissipated becomes equal and so they lose much of their kinetic energy, thus decreasing the Reynolds number to the order of unity. The inner scale $l_o$ is typically of the order of millimeters near the ground while the outer scale $L_o$ can take values in meters. Both quantities increase as one moves vertically away from the ground.

At optical frequencies, the refractive index of the atmosphere is known to depend on the local pressure and temperature through the relation [3, 5]:

$$n_1 = 77.6\left[1 + 7.52 \times 10^{-3}\lambda^{-2}\right]\frac{P}{T} \times 10^{-6} \tag{7.9}$$

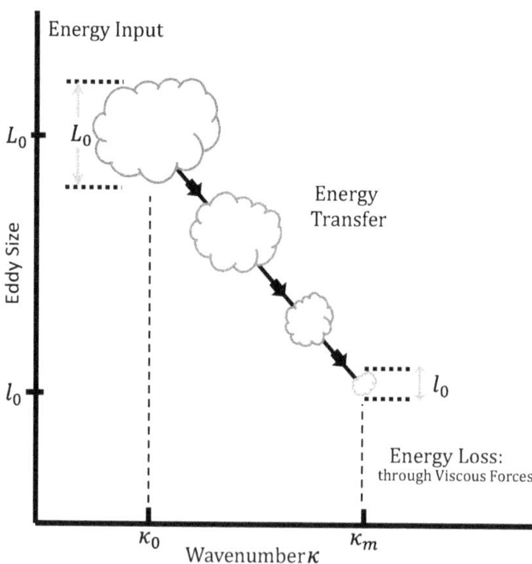


**Figure 7.1.** Energy cascade theory of turbulence: the large sized eddies subdivide into smaller and smaller eddies until they completely disappear. The homogeneous and isotropic turbulence is characterized by eddies with sizes lying between $L_o$ and $l_o$ which denote the outer and inner scales of turbulence respectively.

where $T$ and $P$ denotes the temperature and pressure of the air in Kelvin and millibar respectively. Pressure fluctuations measured at a point with respect to the ground are relatively small and as a result, the temperature fluctuations play a major role in producing refractive index variations.

### 7.2.2 Spatial statistics of refractive index fluctuations

*Correlation function and power-spectral density*
The fluctuating part of refractive index given by $n_1(r)$ is a random process as given by equation (7.6). The random process $n_1(r)$ is usually assumed to be stationary and statistically homogeneous. As a result, the auto-correlation of $n_1(r)$, given by $\Gamma_n(r_1, r_2)$ depends only on the spatial separation $r = r_1 - r_2$ and not on the exact spatial positions $r_1$ and $r_2$ [6].

Therefore we write[1],

$$\Gamma_n(r_1, r_2) = \langle n_1(r_1)n_1(r_2)\rangle$$
$$\Gamma_n(r) = \langle n_1(r_1)n_1(r_1 - r)\rangle. \tag{7.10}$$

The statistical distribution of the size and number of the turbulent eddies is characterized by the spatial power-spectral density (PSD) of $n_1(r)$, denoted as $\Phi_n(\kappa)$. Here $\kappa$ is the wavenumber vector with orthogonal components: $\kappa_x = 2\pi f_x$, $\kappa_y = 2\pi f_y$ and $\kappa_z = 2\pi f_z$ and $|\kappa| = \sqrt{\kappa_x^2 + \kappa_y^2 + \kappa_z^2}$. The auto-correlation function $\Gamma_n(r)$ and the power-spectral density $\Phi_n(\kappa)$ are related by the Weiner–Khinchine theorem (see chapter 2) and form a three-dimensional Fourier transform pair [6]:

$$\Gamma_n(r) = \iiint_{-\infty}^{+\infty} \Phi_n(\kappa)\exp(i\kappa \cdot r)\ d\kappa, \tag{7.11}$$

$$\Phi_n(\kappa) = \left(\frac{1}{2\pi}\right)^3 \iiint_{\infty}^{+\infty} \Gamma_n(r)\exp(-i\kappa \cdot r)\ dr. \tag{7.12}$$

The power-spectral density $\Phi_n(\kappa)$ is considered as a measure of the relative abundance of the turbulent eddies with scales $l_x = 2\pi/\kappa_x$, $l_y = 2\pi/\kappa_y$ and $l_z = 2\pi/\kappa_z$. For the case of homogeneous and isotropic turbulence, the auto-correlation of $n_1(r)$ is spherically symmetric. Hence, $\Gamma_n(r)$ is only a function of the scalar distance $r = |r|$, while $\Phi_n(\kappa)$ is a function of the scalar wavenumber $\kappa$ which is related to the isotropic scale size $l = 2\pi/\kappa$. Both auto-correlation and PSD for homogeneous random processes are even functions [3, 7], allowing us to rewrite equations (7.11) and (7.12) as,

$$\Gamma_n(r) = \frac{4\pi}{r} \int_0^{+\infty} \kappa d\kappa\ \Phi_n(\kappa)\sin(\kappa r), \tag{7.13}$$

---

[1] In the case of any complex stationary random process $x(t)$, the correlation function is defined by $\Gamma_x(t_1, t_2) = \langle x(t_1)x^*(t_2)\rangle$, where the asterisk denotes the complex conjugation.

$$\Phi_n(\kappa) = \frac{1}{2\pi^2\kappa} \int_0^{+\infty} r\,dr\ \Gamma_n(r)\sin(\kappa r). \tag{7.14}$$

*Covariance function*

The spatial covariance function of the refractive index field is defined by the ensemble average,

$$B_n(r_1,\, r_2) = \langle [n_1(r_1) - \langle n_1(r_1)\rangle][n_1(r_2) - \langle n_1(r_2)\rangle]\rangle. \tag{7.15}$$

For statistically homogeneous turbulence, the moments of the refractive index fluctuations will be invariant under spatial translation. Then the mean value $\langle n_1(r)\rangle = m$ has some constant value, independent of the spatial coordinate $r$. The covariance function is now only dependent on the spatial separation $r = r_1 - r_2$,

$$B_n(r) = \langle [n_1(r_1)\, n_1(r_1 + r)]\rangle - |m|^2. \tag{7.16}$$

For refractive index fluctuations, $\langle n_1(r)\rangle = 0$, therefore, the covariance function is same as the auto-correlation function.

*Structure function*

The mean value of the refractive index at a given location $r$ is not a constant and fluctuates over reasonably long length scales. As a result the random process $n_1(r)$ cannot be considered as a homogeneous (or spatially stationary) process. In this case, rather than looking at the random process $n_1(r)$ itself, one can define another random process $n_1(r + r') - n_1(r)$ which behaves in a very similar way to a stationary random process with slowly varying mean. For such processes, the structure function is a valuable statistical descriptor and is defined as [3, 8],

$$D_n(r_1,\, r_2) = \langle [n_1(r_1) - n_1(r_2)]^2\rangle. \tag{7.17}$$

The subtraction process in the structure function removes the effect of slowly varying large scale fluctuations as the difference operation acts like a high pass filter. For a stationary random process, the structure function is related to the auto-correlation function $\Gamma_n(r)$ as,

$$D_n(r) = 2[\Gamma_n(0) - \Gamma_n(r)]. \tag{7.18}$$

Using equation (7.11), the structure function can be written in terms of the PSD as [3]:

$$D_n(r) = 2 \iiint_{-\infty}^{+\infty} [1 - \exp(i\boldsymbol{\kappa} \cdot \boldsymbol{r})]\ \Phi_n(\boldsymbol{\kappa})\ d\boldsymbol{\kappa}. \tag{7.19}$$

For isotropic turbulence, $D_n(r)$ is a function of scalar distance $r$, so we may further obtain the following simplified expression:

$$D_n(r) = 8\pi \int_0^\infty d\kappa\ \Phi_n(\kappa)\left[1 - \frac{\sin(\kappa r)}{\kappa r}\right]\kappa^2. \tag{7.20}$$

We will have occasion to use the structure function in the next chapter when understanding the nature of phase screens generated by the FFT method for simulating atmospheric turbulence.

### 7.2.3 Temporal evolution of the fluctuations

The atmosphere is a dynamically evolving random medium where the distribution of the turbulent eddies changes over millisecond time scales. The temporal fluctuations in the refractive index of the atmosphere play an important role in applications like imaging. As discussed previously, the effect of the time domain fluctuations are modeled using the Taylor's frozen flow hypothesis. This hypothesis treats the entire arrangement of turbulent eddies to be frozen during the measurement interval. The complete frozen air mass only undergoes a horizontal translation in the transverse plane due to the wind as shown in figure 7.2. The transverse velocity of the wind $v$ is assumed to be constant at a given location. Here, the transverse velocity refers to the component of the wind velocity normal to the observation path. The refractive index fluctuation at some time $t_2 > t_1$ is thus related to the fluctuation at $t_1$ by [7],

$$n_1(\mathbf{r},\ t_2) = n_1(\mathbf{r} - v(t_2 - t_1),\ t_1). \tag{7.21}$$

Therefore, the space–time covariance function becomes,

$$\Gamma_n^S(\mathbf{r}_1,\ t_1;\ \mathbf{r}_2,\ t_2) = \langle n_1(\mathbf{r}_1,\ t_1)n_1(\mathbf{r}_2,\ t_2)\rangle, \tag{7.22}$$

$$= \Gamma_n(\mathbf{r}_1 - \mathbf{r}_2 + v(t_2 - t_1)). \tag{7.23}$$

In the above expression, the fluctuations in the wind velocity have been neglected by assuming that the magnitude of velocity fluctuations is much less than the magnitude of the wind velocity. However, this might not always be true for cases when the wind

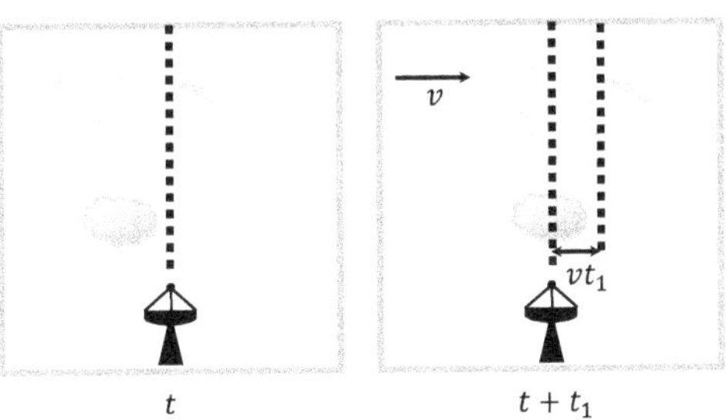


**Figure 7.2.** The Taylor's hypothesis: frozen turbulent eddies translated by the wind's velocity over a time duration $t_1$.

flows near parallel to the line of sight of the optical system. The Taylor hypothesis also breaks down for large time intervals as the turbulent eddies ultimately get rearranged over time.

### 7.2.4 Different models for power-spectral density of the refractive index fluctuations

In this section, we will briefly discuss the important PSD models. As discussed in section 7.2.1, the PSD is only described for the wavenumbers which fall inside the inertial sub-range which is bounded by $l_o$ and $L_o$ where the PSD can be assumed to be isotropic. For wavenumbers lying outside this range, the spectrum is generally considered to be anisotropic and does not have any closed form expression.

1. **Kolmogorov spectrum**

    The Kolmogorov theory predicts the PSD $\Phi_n(\kappa)$ to be of the form [4]:

$$\Phi_n^K(\kappa) = 0.033 \, C_n^2 \, \kappa^{-11/3} \ \text{ for } \ \frac{2\pi}{L_o} < \kappa < \frac{2\pi}{l_o}. \tag{7.24}$$

The quantity $C_n^2$ is called the structure constant of the refractive index fluctuations with units $m^{-2/3}$ which characterizes the strength of the refractive index fluctuations. In the limit $\kappa \to 0$, the expression for the spectrum $\Phi_n(\kappa)$ has a singularity in the form a non-integrable pole. To overcome this difficulty and to incorporate the effect of inner and outer scales, alternate forms of PSD have been proposed.

2. **Tatarskii spectrum**

    As discussed earlier, the turbulent eddies of sizes $l < l_o$ form the viscous dissipation range where the energy is mostly dissipated through viscosity effects in the form of heat. Tatarskii introduced a new model which truncates the spectrum in this dissipation range much more sharply than the $\kappa^{-11/3}$ form given by Kolmogorov [4]. The Tatarskii model for the PSD is given by:

$$\Phi_n^T(\kappa) = 0.033 \, C_n^2 \, \kappa^{-11/3} \exp\left(-\frac{\kappa^2}{\kappa_m^2}\right) \ \text{ for } \ \kappa \gg \frac{2\pi}{L_o}, \tag{7.25}$$

where $\kappa_m = 5.92/l_o$. The Tatarskii spectrum also contains a singularity when $L_o \to \infty$. Both the Kolmogorov and Tatarskii spectra are further modified so that they remain isotropic and finite for $\kappa < 1/L_o$. This brings us to the von Kármán spectrum.

3. **von Kármán spectrum**

    The Kolmogorov spectrum can be modified [4] to include wavenumbers $\kappa < 1/L_o$ as,

$$\Phi_n^V(\kappa) = \frac{0.033 \, C_n^2}{\left(\kappa^2 + \kappa_0^2\right)^{11/6}} \ \text{ for } \ 0 \leqslant \kappa \ll \frac{2\pi}{l_o}, \tag{7.26}$$

where $\kappa_0 = 2\pi/L_o$. Similarly, the Tatarskii spectrum can be modified,

$$\Phi_n^V(\kappa) = \frac{0.033 C_n^2}{\left(\kappa^2 + \kappa_0^2\right)^{11/6}} \exp\left(-\frac{\kappa^2}{\kappa_m^2}\right) \quad \text{for} \ \ 0 \leqslant \kappa < \infty. \tag{7.27}$$

Equations (7.26) and (7.27) are known as the von Kármán spectrum and the modified von Kármán spectrum respectively. Notice that a non-zero $\kappa_0$ results in a finite value of PSD at $\kappa = 0$ while a non-zero $\kappa_m$ rapidly brings the PSD to zero for $\kappa > \kappa_m$. The PSD in equation (7.27) is defined for the entire range of $\kappa$ values, however its form in the range $\kappa < 2\pi/L_o$ is an approximation.

The three spectra given by equations (7.24), (7.26) and (7.27) are shown in figure 7.3. The three distinct stages of turbulence process—the energy injection, inertial range and the energy dissipation are also indicated in the plot. Experimental measurements of the power spectrum of the temperature fluctuations in the atmosphere by Champagne *et al* [9] and by Williams and Paulson [10] showed that at high wavenumbers near $1/l_o$, the temperature spectrum features a small rise (or bump) that causes the spectrum to decrease less rapidly than predicted by the $\kappa^{-11/3}$ law of Oboukhov [11] and Corrsin [12]. Since the refractive index spectrum nominally follows the same spectral law as that of the temperature, this characteristic bump must also feature in its spectrum. However, none of the above spectrum models show this feature. In this regard, starting with first principles, Hill developed a theoretical spectrum [13, 14] for temperature fluctuations in the atmosphere which is in good agreement with the experimentally obtained data. For optical refractive index fluctuations in the atmosphere, the Hill spectrum $\Phi_n^H(\kappa)$ is the solution to a second-order linear homogeneous differential equation [15]:

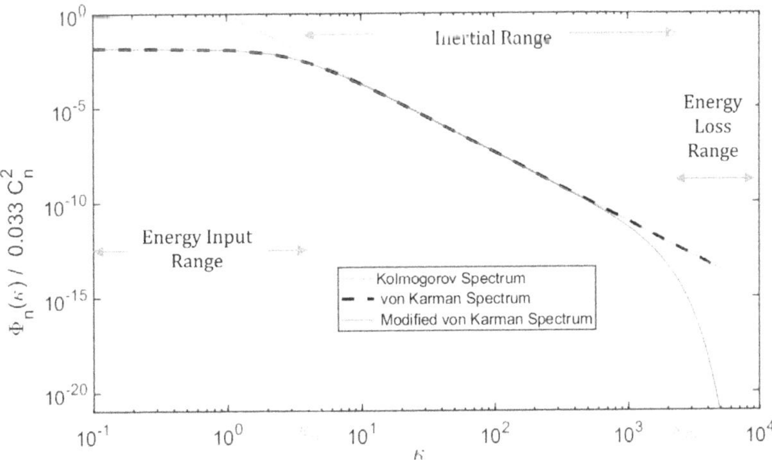


Figure 7.3. The Kolmogorov, von Kármán and modified von Kármán spectrum models for refractive index fluctuations are plotted with respect to wavenumber $\kappa$ (m$^{-1}$) where $l_o = 5$ mm and $L_o = 2$ m.

$$\frac{d}{d\kappa}\left\{\kappa^{14/3}\left[(13.9\kappa\eta)^{3.8} + 1\right]^{-0.175}\frac{d}{d\kappa}\Phi_n^H(\kappa)\right\} = 14.1\kappa^4\eta^{4/3}\Phi_n^H(\kappa), \qquad (7.28)$$

where $\eta = 0.135l_o$. Hill and Clifford showed that the use of Hill spectrum in optical propagation calculations (like the variance of log intensity, the structure function of phase, and phase coherence length) can produce significantly different results than the Tatarskii spectrum[14]. Churnside provided the following analytical approximation to the Hill spectrum [15]:

$$\Phi_n^H(\kappa) \approx \Phi_n^K(\kappa)\left\{\exp(-70.5\kappa^2\eta^2) + 1.45\exp[-0.97(\ln(\kappa\eta) + 1.55)^2]\right\}, \qquad (7.29)$$

where $\Phi_n^K(\kappa)$ is the Kolmogorov spectrum defined in equation (7.24). The above form of the Hill spectrum is useful for numerical integration of quantities that involve the refractive index power spectrum, but is not convenient for pure analytical studies. In this regard, Andrews gave the following approximate analytical form [16] of the Hill spectrum which is more convenient to use:

$$\Phi_n^M(\kappa) = 0.033C_n^2\left[1 + 1.802(\kappa/\kappa_l) - 0.254(\kappa/\kappa_l)^{7/6}\right]\frac{\exp(-\kappa^2/\kappa_l^2)}{(\kappa_0^2 + \kappa^2)^{11/6}}, \qquad (7.30)$$

$$\text{for } 0 \leqslant \kappa < \infty \quad \kappa_l = 3.3/l_o.$$

This approximation is also known as the modified atmospheric spectrum [3].

### 7.2.5 Behavior of turbulence strength: $C_n^2$ models

As seen from all the spectrum models discussed in the previous section, the refractive index structure constant $C_n^2$ plays a very important role in the description of the atmospheric turbulence. In the above description, we have assumed $C_n^2$ to be a constant, representing an averaged quantity. However, in reality $C_n^2$ is a random variable and depends on the temperature instabilities at a given geographical location and time. In general, $C_n^2$ is a seen to be a function of altitude, location and the time of the measurement [4]. Extensive experimental research has been going on to measure the $C_n^2$ profiles at various altitudes and locations resulting in the development of different mathematical models. The $C_n^2$ measurements have been collected at various locations and altitudes at different times of the day and year using optical scintillometers. These devices include a transmitter which emits electromagnetic radiation parallel to the propagation path and a receiver situated at some distance $z$ from the transmitting aperture. The receiver measures the scintillation (or intensity fluctuations) in the beam due to the turbulence present in the atmospheric path. The numerical value of the intensity scintillation is then equated to the theoretical Rytov variance values and an estimate for $C_n^2$ is obtained. The generally observed values for the $C_n^2$ at a height of 2 m above ground is $10^{-17}\ \text{m}^{-2/3} < C_n^2 < 10^{-12}\ \text{m}^{-2/3}$ on a typical sunny day [5]. The important fluctuations in $C_n^2$ can be attributed to the following factors [7].

1. **Temperature fluctuations**: The $C_n^2$ values when measured over a 24 h period show a characteristic diurnal cycle. It has been observed that the $C_n^2$ shows an increasing trend till late afternoon and then starts falling to lower values after sunset. During daytime, when the sunlight is constantly heating up the Earth's surface, the air near the surface becomes hot. This hot air rises up and turbulently mixes with the cooler air, thus increasing the temperature irregularities in the atmosphere that give rise to refractive index fluctuations. Similarly, when the Sun sets and the surface starts to cool down, this turbulent mixing of hot and cool air subsides thereby reducing the refractive index fluctuations in the atmosphere. Minimum values of $C_n^2$ are usually observed at sunrise and sunset times when the surface and air temperatures are similar.

2. **Altitude variation**: Many propagation problems of practical interest involve vertical or slanted beam propagation paths. In these cases, it becomes necessary to know the behavior of $C_n^2$ with altitude. The experimental studies to measure the $C_n^2$ profiles along a vertical path have been undertaken by mounting temperature sensors at various heights, for example, via the use of tower, aircraft or balloon mounted sensors. The $C_n^2$ values are obtained by measuring the $C_T^2$ (temperature fluctuation) values at different heights [4]. It is observed that with increasing altitude $h$, the $C_n^2$ values approximately decrease as per the power law $h^{-4/3}$. Many $C_n^2$ distributions as a function of altitude have been proposed [7], e.g. the Hufnagel–Valley model, the Greenwood model and the Submarine Laser Communication Day (SLC-Day) model. The commonly used Hufnagel–Valley model approximates the $C_n^2$ profile as:

$$C_n^2(h) = 5.94 \times 10^{-53}(v/27)^2 h^{10} \exp(-h/100) \\ + 2.7 \times 10^{-16} \exp(-h/1500) + A \exp(-h/100), \tag{7.31}$$

where, $A$ sets the turbulence strength near ground with a typical value of $A = 1.7 \times 10^{-14}$ m$^{-2/3}$. The parameter $v$ is the high altitude wind speed with commonly used value for this parameter is $v = 21$ ms$^{-1}$.

Description of laser beam propagation in atmospheric turbulence thus needs careful attention to various models and parameters as discussed here. It is important to understand the approximations or empirical nature associated with these models when trying to simulate laser beam propagation through the atmosphere at a specific geographical location. Empirical measurements of years-long variations in quantities such as $C_n^2$, temperature, pressure, etc, at specific locations may therefore be required before prediction of experimental behavior of laser beam profiles can be made satisfactorily using numerical simulations.

## 7.3 Classical perturbation methods

The classical perturbation techniques include the Born approximation and Rytov approximation. Both of these perturbation methods are restricted to weak turbulence regimes. These approaches do take into account diffraction effects unlike the geometrical optics model. We will describe these historically important methods in some detail in the following discussion.

### 7.3.1 Born approximation

The first classical perturbation technique that we will discuss is the Born approximation (figure 7.4). This technique was first used in quantum mechanics to study scattering problems. Light experiences scattering on interacting with the turbulent eddies present in the atmosphere. However, the term $n_1(r)$ which represents the fluctuations in the refractive index about its mean value is very small, in fact it is of the order of a few parts in $10^5$–$10^6$. It is reasonable to expect that the light beam will propagate through the atmosphere very much like it does in free space but with additional small variations as a result of the scattering. In the weak scattering limit, the Born approximation states that the field $U(r)$ can be written as a sum of terms of the form:

$$U(r) = U_0(r) + \nu U_1(r) + \nu^2 U_2(r) + \cdots \tag{7.32}$$

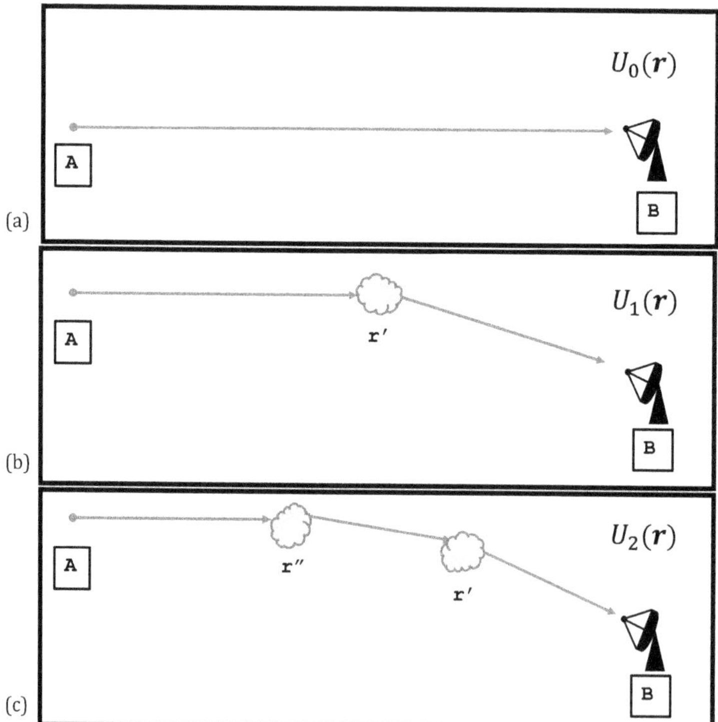


**Figure 7.4.** Schematic diagram showing the first three terms in the Born approximation. The transmitter at point A emits plane wave which travels through the medium to the receiver situated at point B. The first term in the Born approximation $U_0(r)$ represents the unperturbed wave as shown in (a). The second term $U_1(r)$ and the third term $U_2(r)$ represent the single and double scattered radiation respectively. These cases are represented in (b) and (c) respectively.

where $\nu$ is a dummy variable that shows the order of smallness of the terms in the expansion. The first term $U_0(r)$ represents the unscattered portion of the field or in other words, the field which would be obtained on free space propagation. The successive terms represent the scattered portion of the field for example, $U_1(r)$ corresponds to single scattering of the incident field by a turbulent eddy and $U_2(r)$ represents double scattering and so on. If the magnitude of each successive term is sufficiently smaller than its immediate preceding term, i.e. $|U_{m+1}| \ll |U_m|$, then the series will converge and a finite number of terms will be able to completely describe the solution. However, this may not be true for all propagation problems, specifically when the total integrated refractive index fluctuations over the propagation path becomes large. This can happen due to either an increase in $n_1(r)$ or an increase in the length of the propagation path. In such cases, multiple scattering effects become dominant and the basic condition of weak scattering as required in the Born approximation is violated.

The average value of refractive index in the atmosphere is $\langle n(r) \rangle \approx 1$, so using equation (7.6), we can write

$$n^2(r) = (1 + \nu n_1(r))^2 \approx 1 + 2\nu n_1(r), \tag{7.33}$$

where we have neglected the second-order term $n_1^2(r)$ as it is very small in magnitude. Substituting for $U(r)$ and $n^2(r)$ in equation (7.8) from equations (7.32) and (7.33) respectively and equating terms corresponding to equal powers of $\nu$ to zero yields the following system of equations:

$$\nu^0: \ \nabla^2 U_0(r) + k^2 U_0(r) = 0, \tag{7.34}$$

$$\nu^1: \ \nabla^2 U_1(r) + k^2 U_1(r) = -2k^2 n_1(r) U_0(r), \tag{7.35}$$

$$\nu^2: \ \nabla^2 U_2(r) + k^2 U_2(r) = -2k^2 n_1(r) U_1(r), \tag{7.36}$$

and so on for higher values of $\nu$. An important outcome of the Born approximation is that we have broken down the original stochastic Helmholtz equation which was a homogeneous partial differential equation with random coefficients into a system of non-homogeneous partial differential equations with constant coefficients and source terms These equations can be solved using the Green function method. The partial differential equation (equation (7.35)) corresponding to the first-order Born approximation term $U_1(r)$ is usually solved. Therefore, the Born approximation is also sometimes called the *single scattering approximation* as we are looking for solutions of the form $U_B^{(1)}(r) = U_0(r) + U_1(r)$, neglecting all higher order scattering terms The solution of equation (7.35) is equal to the convolution of the free space Green function $G(r)$ with the source term $-2k^2 n_1(r) U_0(r)$. The Green function in free space is given by:

$$G(r) = \frac{1}{4\pi} \frac{e^{ik|r|}}{|r|}. \tag{7.37}$$

Then, the solution for $U_1(r)$ can be written as:

$$U_1(r) = \frac{1}{4\pi} \int_V \frac{e^{ik|r-r'|}}{|r - r'|}[2k^2 n_1(r')U_0(r')]d^3r', \tag{7.38}$$

where $V$ is the scattering volume. At any position $r'$ inside the scattering volume $V$, the incident field $U_0(r')$ interacts with the turbulent eddies present in the atmosphere and gets scattered in the form of secondary spherical waves which are collected at the receiver situated at $r$. Hence, the first correction term of the Born approximation $U_1(r)$ is just the summation of these scattered spherical waves inside the scattering volume where the strength of these waves is given by the product of the incident field $U_0(r')$ and the refractive index fluctuations $n_1(r')$. An important point to note is that by definition, it is assumed that $\langle n_1(r)\rangle = 0$, then equation (7.38), implies that the ensemble average or mean of the first Born approximation term also vanishes $\langle U_1(r)\rangle = 0$. Higher order perturbation terms $U_2(r)$, $U_3(r)$, etc can be solved in the same manner, however we will not discuss these solutions here.

*Paraxial approximation*
The solution for $U_1(r)$ as given by equation (7.38) can be converted into a much simpler form by making use of certain approximations. We begin by assuming that the backscattering of the incident wave can be neglected completely. This is a direct consequence of the weak scattering approximation where the incident wave is assumed to be scattered only in the forward direction by the weak turbulent eddies. The smallest sized eddies are responsible for the largest angle of scattering. Therefore, considering the inner scale of turbulence to be equal to $l_o$, the maximum scattering angle would be of the order of $\theta = \lambda/l_o$. The smallest eddies in the atmosphere are of the order of few millimeters, so that $\lambda \ll l_o$. The scattering angle for typical values of $\lambda = 0.6\,\mu m$ and $l_o = 2\,mm$ comes out to be equal to $\theta = 3 \times 10^{-4}$ radians, which is small enough. Figure 7.5 shows the scattering of an incident plane wave by turbulent eddies of various spatial sizes $l$. In order for the scattered waves from any turbulent eddy to reach the receiver situated at $P$, the maximum lateral displacement of that turbulent eddy from the propagation axis should follow the relation:

$$\rho \leqslant \frac{\lambda z}{l_o}. \tag{7.39}$$

This condition limits the lateral extent of the turbulent eddies which contribute to the optical field received at $P$. In other words, this means that the turbulent eddies which are far from the propagation axis result in scattered waves which never reach the receiver. Therefore, the received optical field at $P$ can be visualized to contain waves scattered in a narrow cone about the wave propagation direction. Typically the longitudinal distance $z$ from the scatterer to receiver is much larger than the transverse displacement $\rho$ from the $z$-axis, i.e. $z \gg \rho$. This allows us to use *paraxial approximation* to expand the exponential term in the Green function of the equation (7.37) in a binomial series:

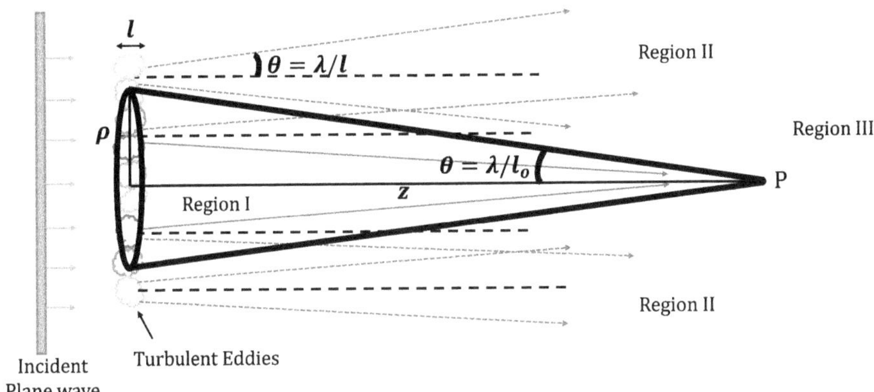


**Figure 7.5.** The figure shows the forward scattering of an incident plane wave by the turbulent eddies of spatial sizes 'l'. The eddies scatter the incident radiation at narrow angles given by $\theta = \lambda/l$. The major contribution to the optical field at point $P$ is from the radiation scattered from eddies situated near to the $z$-axis, as represented by region I. The contribution from the forward-scattered radiation from region II to the field at point $P$ is very small. In region III, only back-scattered radiation reaches $P$.

$$k|\mathbf{r}| = kz\sqrt{1 + \left(\frac{\rho}{z}\right)^2} = kz\left[1 + \frac{1}{2}\left(\frac{\rho}{z}\right)^2 - \frac{1}{8}\left(\frac{\rho}{z}\right)^4 + \cdots\right]. \tag{7.40}$$

The first term in this expansion gives the phase evolution of a wave in free space and is always retained in the expansion. The higher order terms are retained if they are significant enough to cause a phase change of more than $2\pi$. The second term is of the order of $\pi\rho^2/\lambda z$ where the maximum value of $\rho$ is set by equation (7.39) and $\sqrt{\lambda z}$ is called as the *Fresnel length* of the system. For a typical propagation distance of $z = 1$ km, $\lambda = 0.6$ μm and $l_o = 2$ mm, the second term remains large and is retained in the expansion. However, the third term is of the order of $\pi\rho^4/4\lambda z^3$ whose value is much less than 1 radian for the above parameters and is therefore inconsequential to us and can be safely dropped. Similarly, the denominator in the Green function can be simply approximated as $|\mathbf{r}| \approx z$. This allows us to write solution $U_1(\mathbf{r})$ in a simple form:

$$U_1(\mathbf{r}) = \frac{k^2}{2\pi}\int_V \exp\left(ik\left[(z - z') + \frac{(\rho - \rho')^2}{2(z - z')}\right]\right)\frac{n_1(\mathbf{r}')U_0(\mathbf{r}')}{(z - z')}d^3\mathbf{r}'. \tag{7.41}$$

The volume integration in the above equation can be separated into three regions as shown in figure 7.5. Region I is the weakly scattered waves from the turbulent eddies contained in a narrow cone of angle $\theta = \lambda/l_o$ as explained earlier. It represents the most significant contribution to the optical field at point $P$. Due to the condition that wavelength $\lambda$ is much smaller compared to the inner scale $l_o$ of turbulence, the contribution from regions II and III which represents large angle scattering and backscattering respectively is negligible. The contribution from region II and region

III to the integral in equation (7.41) can be shown to be of the order of $\approx n_1^2/16$ small using the method of stationary phase. Large angle scattering is also not of utmost importance in understanding the laser beam profile after propagating a distance of a few km from the laser source through turbulence. The light received by a detector or receiver at such large distances mainly represents small spatial frequencies associated with scattered light.

*Depolarization of light*
Let us revisit equation (7.1) and pay some attention to the depolarization term which we have ignored so far. The scattering in the atmosphere can give rise to subtle changes in the polarization of the light beam. As discussed earlier, for long range line of sight propagation the scattering angles involved are usually very small and so significant changes in the polarization solely due to the refractive index inhomogeneities are not observed. In such cases, the major depolarization of the light beam is caused due to wide-angle scattering from aerosols, pollutants, dust particles, rain drops, etc, present in the atmosphere. However, there might be some wide-angle scattering in the troposphere which can give rise to significant changes in the polarization of the beam for propagation beyond the horizon. Two different approaches based on diffraction theory and geometrical optics have been developed for describing the depolarization of light by atmosphere. Here, we would look into the calculation first provided by Tatarskii using the Born approximation [17].

Let us assume that a linearly polarized plane wave is traveling along the $z$-axis and we want to calculate by how much its electric field vector gets rotated on propagation through the atmosphere. Assuming a sinusoidal time dependence for the electric field (see equation (7.2)), we can write equation (7.1) as:

$$\nabla^2 E_o(r) + \nabla(E_o(r) \cdot \nabla \ln \varepsilon) + k^2 n^2(r) E_o(r) = 0. \tag{7.42}$$

Using the Born approximation, we write $E_o(r)$ as:

$$E_o(r) = E_0(r) + \nu E_1(r), \tag{7.43}$$

where $E_0(r)$ is the incident field and $E_1(r)$ represents the scattered field. Let the incident field be $x$-polarized, then $E_0(r) = E_x^0(r)\hat{x}$ but the scattered field also contains $y$-polarized components due to depolarization, therefore $E_1(r) = E_x^1(r)\hat{x} + E_y^1(r)\hat{y}$. Substituting for the quantities $E_o(r)$, $n^2(r)$ and $\ln \varepsilon = 2 \ln n$ in equation (7.42) to collect equal powers of $\nu$ :

$$\nu^0: \ \nabla^2 E_0(r) + k^2 E_0(r) = 0, \tag{7.44}$$

$$\nu^1: \ \nabla^2 E_1(r) + k^2 E_1(r) = -2k^2 n_1(r) E_0(r) - 2\nabla(E_0(r). \ \nabla n_1(r)). \tag{7.45}$$

Here, $\ln(1 + n_1(r))$ has been expanded in a power series in $n_1(r)$. The depolarized component of the scattered wave is given by:

$$\nabla^2 E_y^1 + k^2 E_y^1 = -2E_x^0(z)\frac{\partial^2 n_1}{\partial y \partial x}. \tag{7.46}$$

Therefore, the solution of the depolarized component can be obtained using the Green function and is given by :

$$E_y^1(r) = \frac{1}{4\pi} \int_V \frac{e^{ik|r-r'|}}{|r-r'|}\left[2E_x^0(z')\frac{\partial^2 n_1(r')}{\partial y'\partial x'}\right]d^3r' \tag{7.47}$$

The mean squared polarization fluctuation is defined as the average intensity of the depolarized component divided by the intensity of the incident plane wave:

$$\langle M^2\rangle = \left\langle \left|\frac{E_y^1(r)}{E_x^0(r)}\right|^2 \right\rangle. \tag{7.48}$$

Substituting for $E_y^1(r)$ in this expression, we can write:

$$\langle M^2\rangle = \frac{1}{4\pi^2} \int_V d^3r' \left(\frac{e^{ik|r-r'|}}{|r-r'|}\frac{E_x^0(z')}{E_x^0(r)}\right)$$
$$\times \int_V d^3r'' \left(\frac{e^{ik|r-r''|}}{|r-r''|}\frac{E_x^0(z'')}{E_x^0(r'')}\right)^* \left\langle \frac{\partial^2 n_1(r')}{\partial y'\partial x'}\frac{\partial^2 n_1(r'')}{\partial y''\partial x''}\right\rangle. \tag{7.49}$$

The quantity in the angle brackets which gives the ensemble average of the double gradient product can be written in terms of the refractive index spectrum [17, 18]. When the refractive index inhomogeneities are assumed to follow the Tatarskii spectrum with an inner scale cutoff wavenumber $\kappa_m$, then for the case of a point receiver, $\langle M^2\rangle$ is given by [17, 18]:

$$\langle M^2\rangle = 0.070 C_n^2 \kappa_m^{7/3} k^{-2} L, \tag{7.50}$$

where $L$ denotes the total path length. Assuming $l_o = 2$ mm, $\lambda = 0.6\ \mu$m, $L = 1$ km and $C_n^2 = 10^{-13}$, we get $\langle M^2\rangle = 8 \times 10^{-18}$ which is far too small to be detected by the conventional optical systems [19, 20]

## 7.3.2 Rytov approximation

In the stochastic Helmholtz equation (7.8) the second term is a product of the stochastic function $n(r, t)$ with the scalar field $U(r)$. This makes it very difficult to solve this equation as the coefficients of the required solution $U(r)$ are random functions of space and time. Rytov transformation provides a means to separate these two functions and convert this equation into the *Riccati equation*. In this newer form, the randomly varying function $n(r, t)$ appears as an additional source term. This transformation was first described by Rytov in 1937, during his investigations on diffraction of light by ultrasonic waves [21]. Later on, Obukhov used this transformation to study wave propagation in a random medium [22]. Rytov's transformation involves substituting the scalar field $U(r)$ as

$$U(r) = U_0(r)\exp(\psi(r)), \tag{7.51}$$

into the Helmholtz equation where $U_0(r)$ represents the unperturbed field in the absence of turbulence. Here, the function $\psi(r) = \chi + iS$ is a complex quantity which represents the effect of the turbulence on the beam. This substitution leads to an equation of the form:

$$\nabla^2 U_0(r) + 2[\nabla U_0(r) \cdot \nabla \psi(r)] + U_0(r)[\nabla^2 \psi(r) + (\nabla \psi(r))^2] + k^2(1 + 2n_1(r))U_0 = 0. \tag{7.52}$$

As the field $U_0(r)$ is the free space field, therefore, it satisfies the homogeneous wave equation and we have

$$\nabla^2 U_0(r) + k^2 U_0(r) = 0, \tag{7.53}$$

which can be substituted in the above equation. Let us now assume that the field $U_0(r)$ can be further expressed as an exponential of a function $\psi_0(r)$ such that, $U_0(r) = \exp(\psi_0(r))$, then the total field $U(r)$ is simply given by

$$U(r) = \exp(\psi_0(r) + \psi(r)). \tag{7.54}$$

Therefore, the complex function $\psi(r)$ satisfies the equation:

$$\nabla^2 \psi(r) + [\nabla \psi(r)]^2 + 2[\nabla \psi_0 \cdot \nabla \psi(r)] + 2k^2 n_1(r) = 0. \tag{7.55}$$

It can be observed that the required solution $\psi(r)$ of this non-linear partial differential equation has been separated from the stochastic function $n_1(r)$.

Tatarskii made extensive use of the Rytov transformation in his seminal works on wave propagation through turbulence [2, 23] and described a perturbative technique known as the *Rytov approximation* or *method of smooth perturbations* for such problems. In the weak scattering limit, the phase function $\psi(r)$ is written as a sum of terms of form:

$$\psi(r) = \nu \psi_1(r) + \nu^2 \psi_2(r) + \cdots \tag{7.56}$$

where $\nu$ is a dummy variable that shows the order of smallness of the terms in the expansion. If we compare the current expression for $U(r)$ with the Born approximation expression (given by equation (7.32)), we notice a major difference between the two treatments. In the Born approximation, the scalar field $U(r)$ is represented as an additive perturbation series where the field contributions due to the scattering are added to the initial unperturbed wave field. However, the Rytov approximation involves a multiplicative perturbation series.

We can substitute the expansion of $\psi(r)$ and $n^2(r)$ from equations (7.56) and (7.33) into the above non-linear equation (7.55). The terms containing equal powers of $\nu$ can be collected to write the following equations:

$$\nu^0: \ \nabla^2 \psi_0(r) + [\nabla \psi_0(r)]^2 + k^2 = 0, \tag{7.57}$$

$$\nu^1: \ \nabla^2 \psi_1(r) + 2[\nabla \psi_0(r) \cdot \nabla \psi_1(r)] = -2k^2 n_1(r), \tag{7.58}$$

$$\nu^2: \quad \nabla^2\psi_2(r) + 2[\nabla\psi_0(r) \cdot \nabla\psi_2(r)] = -\nabla\psi_1(r) \cdot \nabla\psi_1(r). \tag{7.59}$$

The equation corresponding to $\nu^0$ gives the free space propagation of the wave. The equations corresponding to higher powers of $\nu > 1$ all have a general form:

$$\nabla^2\psi_\nu(r) + 2[\nabla\psi_0(r) \cdot \nabla\psi_\nu(r)] = -\sum_{j=1}^{\nu-1} \nabla\psi_j(r) \cdot \nabla\psi_{\nu-j}(r) = f_\nu(r). \tag{7.60}$$

The usual practice is to solve only for the first-order term. This solution is known as the basic Rytov solution and is used widely to describe beam propagation through random medium. The equations for the higher order terms can be solved if $f_\nu(r)$ is known, which itself depends on the lower order solutions. The series for $\psi(r)$ would converge if the successive terms $\psi_{m+1}$ are smaller than their preceding terms $\psi_m$.

The equation for the first-order term $\psi_1(r)$ depends only on the function $\psi_0(r)$ and the refractive index variation $n_1(r)$. Let us assume the function $\psi_1(r)$ to be in the form:

$$\psi_1(r) = W(r)\exp(-\psi_0(r)). \tag{7.61}$$

This simplifies equation (7.58) to:

$$\nabla^2 W(r) - W(\nabla^2\psi_0(r) + [\nabla\psi_0(r)]^2) = -2k^2 n_1(r)e^{\psi_0}. \tag{7.62}$$

The term inside the brackets on the left-hand side of this equation can be substituted from equation (7.57) and we get:

$$\nabla^2 W(r) + k^2 W(r) = -2k^2 n_1 e^{\psi_0(r)}. \tag{7.63}$$

This equation is solvable using the Green function method as explained in the previous section. Therefore we can write the solution for $W(r)$ as:

$$W(r) = \frac{1}{4\pi} \int_V \frac{e^{ik|r-r'|}}{|r-r'|}\left[2k^2 n_1(r')e^{\psi_0(r')}\right] d^3r'. \tag{7.64}$$

The basic Rytov solution is obtained as:

$$\psi_1(r) = \frac{1}{4\pi} \int_V \frac{e^{ik|r-r'|}}{|r-r'|}\left[2k^2 n_1(r')e^{\psi_0(r')-\psi_0(r)}\right] d^3r', \tag{7.65}$$

or using the relation $U_0(r) = \exp(\psi_0(r))$, one can write

$$\psi_1(r) = \frac{1}{4\pi} \int_V \frac{e^{ik|r-r'|}}{|r-r'|}\left[2k^2 n_1(r')\frac{U_0(r')}{U_0(r)}\right] d^3r'. \tag{7.66}$$

On comparing this relation with equation (7.38), we can see that

$$\psi_1(r) = \frac{U_1(r)}{U_0(r)}, \tag{7.67}$$

where $U_1(r)$ is the first-order scattering term of the Born approximation. The basic Rytov solution can thus be written as:

$$U_R^{(1)}(r) = U_0(r)\exp(\psi_1(r)) = U_0(r)\exp[U_1(r)/U_0(r)]. \tag{7.68}$$

In the limit of weak scattering where $U_1(r) \ll U_0(r)$, the exponential in the above equation can be expanded in a series, which is equal to the first-order Born approximation solution:

$$U_R^{(1)}(r) = U_0(r)\left[1 + \frac{U_1(r)}{U_0(r)}\right] = U_0(r) + U_1(r) = U_B^{(1)}(r). \tag{7.69}$$

It was further shown by Yura [24, 25] that in the weak scattering limit, the solution of the second-order Rytov approximation term $\psi_2(r)$, can be written as a combination of the first- and second-order Born approximation terms:

$$\psi_2(r) = \frac{U_2(r)}{U_0(r)} - \frac{1}{2}\left[\frac{U_1(r)}{U_0(r)}\right]^2. \tag{7.70}$$

This suggests an interesting relationship between the two perturbation methods. Let us try to understand this more by looking at the fourth-order solutions obtained from both the Rytov and Born methods:

$$U_R^{(4)} = U_0(r)\exp\left[\nu\psi_1(r) + \nu^2\psi_2(r) + \nu^3\psi_3(r) + \nu^4\psi_4(r)\right], \tag{7.71}$$

$$U_B^{(4)} = U_0(r)\left[1 + \nu\Phi_1(r) + \nu^2\Phi_2(r) + \nu^3\Phi_3(r) + \nu^4\Phi_4(r)\right], \tag{7.72}$$

where $\Phi_m = U_m(r)/U_0(r)$ gives the *normalized Born perturbation terms* and $\nu$ is a dummy variable which gives the order of smallness of the terms. On equating the two solutions we get:

$$\begin{aligned}
&\nu\psi_1(r) + \nu^2\psi_2(r) + \nu^3\psi_3(r) + \nu^4\psi_4(r) \\
&= \ln\left[1 + \nu\Phi_1(r) + \nu^2\Phi_2(r) + \nu^3\Phi_3(r) + \nu^4\Phi_4(r)\right].
\end{aligned} \tag{7.73}$$

As the magnitude of the Born approximation terms $U_m(r)$ are assumed to be smaller than the unscattered term $U_0(r)$, therefore the natural logarithm in equation (7.73) can be expanded in a Maclaurin series using the relation:

$$\ln(1 + x) = x - \frac{x^2}{2} + \frac{x^3}{3} - \frac{x^4}{4} + \cdots \tag{7.74}$$

By equating the equal powers of $\nu$ on both sides of the equation, the first four Rytov perturbation terms can be approximated as:

$$\psi_1(r) = \Phi_1(r), \tag{7.75}$$

$$\psi_2(r) = \Phi_2(r) - \frac{1}{2}\Phi_1^2(r), \tag{7.76}$$

$$\psi_3(r) = \Phi_3(r) - \Phi_1(r)\Phi_2(r) + \frac{1}{3}\Phi_1^3(r), \tag{7.77}$$

$$\psi_4(r) = \Phi_4(r) - \frac{1}{2}\Phi_2^2(r) - \frac{1}{4}\Phi_1^4(r) - \Phi_1(r)\Phi_3(r) + \Phi_1^2(r)\Phi_2(r). \tag{7.78}$$

This analysis suggests that the Rytov perturbation terms can be approximately expressed by a sum of different perturbation terms of the Born series. However, there are some important differences between the two treatments which become apparent when the phase fluctuations in the beam are large. In such cases, the exponential term in the Rytov method cannot be approximated by Born perturbation terms as now $U_m(r)$ might not be very small compared to the unscattered term $U_0(r)$. In such cases, the Rytov method provides a superior description and is preferred over the Born series.

*Amplitude and phase fluctuations*
We have established the perturbative solutions using both the Born and Rytov approximation methods. We are now in a position to investigate amplitude and phase fluctuations experienced by the incident optical field on its interaction with the turbulent atmosphere. Let us begin by assuming that the field after experiencing turbulence $U(r)$ is represented by amplitude $A$ and phase $s$ while the incident (or the free space) field $U_0(r)$ has amplitude $A_0$ and phase $s_0$, that is:

$$U(r) = A\exp(is), \tag{7.79}$$

$$U_0(r) = A_0\exp(is_0). \tag{7.80}$$

Then,

$$\frac{U(r)}{U_0(r)} = \frac{A}{A_0}\exp[i(s - s_0)]. \tag{7.81}$$

Let $U(r)$ be given by the basic Rytov solution (see equation (7.68)), then one can write:

$$\frac{U(r)}{U_0(r)} = \exp(\psi_1). \tag{7.82}$$

Equating the above two equations and taking the logarithm on both sides gives:

$$\psi_1 = \ln\left[\frac{A}{A_0}\right] + i(s - s_0). \tag{7.83}$$

The log-amplitude fluctuations $\chi$ and the phase fluctuation $S$ are defined as :

$$\chi = \ln\left[\frac{A}{A_0}\right], \tag{7.84}$$

$$S = s - s_0. \qquad (7.85)$$

Therefore, one can write

$$\psi_1 = \chi + iS. \qquad (7.86)$$

Thus we see that the log-amplitude fluctuations and phase fluctuations are equal to the real and imaginary part of $\psi_1(r)$. By invoking the paraxial approximation described in the last section, one can write $\chi$ and $S$ as:

$$\begin{bmatrix} \chi(r) \\ S(r) \end{bmatrix} = \begin{bmatrix} \mathrm{Re} \\ \mathrm{Im} \end{bmatrix} \left\{ \frac{k^2}{2\pi} \int_V \exp\left( ik\left[ (z - z') + \frac{(\rho - \rho')^2}{2(z - z')} \right] \right) \frac{n_1(r')}{(z - z')} \frac{U_0(r')}{U_0(r)} d^3r' \right\}. \qquad (7.87)$$

For the present discussion, we would consider a unit amplitude plane wave propagating in the $+z$ direction as the original incident (or unperturbed) field, i.e. $U_0(r) = \exp(ikz)$. This reduces the above equation to the form:

$$\begin{bmatrix} \chi(r) \\ S(r) \end{bmatrix} = \begin{bmatrix} \mathrm{Re} \\ \mathrm{Im} \end{bmatrix} \left\{ \frac{k^2}{2\pi} \int_V \exp\left[ \frac{ik(\rho - \rho')^2}{2(z - z')} \right] \frac{n_1(r')}{(z - z')} d^3r' \right\}. \qquad (7.88)$$

The expressions for log-amplitude and phase fluctuations for the other forms of incident fields (spherical, Gaussian) can be found in a similar fashion. Also notice that though we are considering log-amplitude fluctuation, in the weak scatter limit, this quantity is identical to the normalized amplitude fluctuation: $\chi = \ln\left[ \frac{A}{A_0} \right] \simeq \frac{A}{A_0}$.

The amplitude and phase fluctuation in the beam increases as the beam propagates along the $z$-axis and experiences more turbulence. For a description of this random process, one needs to consider the averaged or statistical quantities. From the previous discussion, we know that $\langle n_1(r) \rangle = 0$ and so the first-order moments for the log-amplitude and phase fluctuations also vanish, that is $\langle \chi(r) \rangle = \langle S(r) \rangle = 0$. This is true because in the current treatment, only the first-order perturbation term $\psi_1(r)$ has been used for calculation of the total field $U(r)$. If higher order perturbation terms are also included, then $\langle \chi(r) \rangle \neq \langle S(r) \rangle \neq 0$ but still these mean values would be inconsequentially small. In order to calculate the higher order statistical moments, it is useful to derive spectral representations of the amplitude and phase fluctuations first.

*Spectral representation of amplitude and phase fluctuations*
Consider the log-amplitude and phase fluctuations as given in last section:

$$\begin{bmatrix} \chi(r) \\ S(r) \end{bmatrix} = \frac{k^2}{2\pi} \int_V \frac{n_1(r')}{(z - z')} \begin{bmatrix} \cos \\ \sin \end{bmatrix} \left[ \frac{k(\rho - \rho')^2}{2(z - z')} \right] d^3r'. \qquad (7.89)$$

The refractive index function $n_1(r')$ can be expressed as a two-dimensional Fourier–Stieltjes integral:

$$n_1(r') = \int d\nu(\kappa, z') \exp(i\kappa \cdot \rho'), \tag{7.90}$$

where $\kappa$ represents the three-dimensional wave-vector with components $(\kappa_x, \kappa_y, \kappa_z = 0)$. The function $\nu(\kappa, z')$ is a random function of propagation distance $z$ and it represents the amplitude of the refractive index fluctuations in a transverse plane perpendicular to the $z$-axis. On substituting this form of $n_1(r')$ into equation (7.89) and expanding $d^3r' = d^2\rho'dz'$, we can solve for $\chi(r)$,[2]

$$\chi(r) = \frac{k^2}{2\pi} \int \int_0^z dz' \frac{d\nu(\kappa, z')}{(z - z')} \int d^2\rho' \exp(i\kappa \cdot \rho') \cos\left[\frac{k(\rho - \rho')^2}{2(z - z')}\right]. \tag{7.91}$$

Let the integral over $\rho'$ be denoted by $W$, this integral can be solved by substituting $\rho' = \rho'' + \rho$, therefore one can write

$$W = \exp(i\kappa \cdot \rho) \int d^2\rho'' \exp(i\kappa \cdot \rho'') \cos\left[\frac{k(\rho'')^2}{2(z - z')}\right], \tag{7.92}$$

$$W = \exp(i\kappa \cdot \rho) \int_0^\infty \rho'' d\rho'' \cos\left[\frac{k(\rho'')^2}{2(z - z')}\right] \int_0^{2\pi} d\theta \exp\left[iK\rho'' \cos(\theta - \alpha)\right], \tag{7.93}$$

where the polar coordinates $\rho'' = (\rho'', \theta)$ and $\kappa = (K, \alpha)$ are used. The inner $\theta$ integral can be solved using the following Bessel identity:

$$J_o(K\rho) = \frac{1}{2\pi} \int_0^{2\pi} \exp(i\kappa\rho \cos(\theta)), \tag{7.94}$$

where $J_o(\kappa\rho)$ is the Bessel function of first kind with order zero. This gives us the expression:

$$W = 2\pi \exp(i\kappa \cdot \rho) \int_0^\infty \rho'' d\rho'' J_0(\kappa\rho'') \cos\left[\frac{k(\rho'')^2}{2(z - z')}\right]. \tag{7.95}$$

This integral is equal to the Hankel transform of the cosine term and so equation (7.95) may be written as:

$$W = \frac{2\pi(z - z')}{k} \exp(i\kappa \cdot \rho) \sin\left[\frac{\kappa^2(z - z')}{2k}\right]. \tag{7.96}$$

Therefore, one can write amplitude and phase fluctuations as:

---

[2] Similar steps can be carried out to obtain the expression for phase fluctuations $S(r)$.

$$\begin{bmatrix} \chi(r) \\ S(r) \end{bmatrix} = \int \exp(i\kappa \cdot \rho) \left\{ k \int_0^z dz' d\nu(\kappa, z') \begin{bmatrix} \sin \\ \cos \end{bmatrix} \left[ \frac{\kappa^2(z - z')}{2k} \right] \right\}. \tag{7.97}$$

On comparing equations (7.90) and (7.97), a similarity between the two expressions can be observed. The term in the curly brackets in equation (7.97) can be interpreted as the random spectral amplitudes of the amplitude and phase fluctuations and it plays the same role as that played by the random function $\nu(\kappa, z')$ in equation (7.90).

*Covariance function for the amplitude and phase fluctuations*
We know from previous discussions that $\langle \chi(r) \rangle = \langle S(r) \rangle = 0$. It is commonly assumed that the amplitude and phase fluctuations are statistically homogeneous (stationary) in the given $z$ plane. Therefore, one can evaluate the two-dimensional covariance function for the amplitude and phase fluctuations, $B_\chi(\rho)$ and $B_S(\rho)$ respectively:

$$B_\chi(\rho) = \langle \chi(\rho_1 + \rho, z) \chi^*(\rho_1, z) \rangle, \tag{7.98}$$

$$B_S(\rho) = \langle S(\rho_1 + \rho, z) S^*(\rho_1, z) \rangle. \tag{7.99}$$

Substituting equation (7.97) into equation (7.98), the expression for $B_\chi(\rho)$ is obtained as[3]:

$$\begin{aligned} B_\chi(\rho) = k^2 \int \int e^{i\kappa \cdot (\rho_1 + \rho) - i\kappa' \cdot \rho_1} \int_0^z dz' \int_0^z dz'' \sin\left( \frac{\kappa^2(z - z')}{2k} \right) \\ \sin\left( \frac{\kappa'^2(z - z'')}{2k} \right) \quad \cdot \langle d\nu(\kappa, z') d\nu(\kappa', z'') \rangle. \end{aligned} \tag{7.100}$$

The ensemble averages of any statistically homogeneous random process are independent of the location at which they are being computed. That is, if we translate our sensors to any other position in the field, and perform the same operation, then the same average properties of the random process should be expected. Therefore, the random amplitude term in the angular brackets should satisfy the relation:

$$\langle d\nu(\kappa, z') d\nu(\kappa', z'') \rangle = \delta(\kappa - \kappa') F_n(\kappa', z' - z'') d^2\kappa d^2\kappa'. \tag{7.101}$$

The function $F_n(\kappa', z' - z'')$ is the two-dimensional spectral density of the refractive index fluctuations defined by:

$$F_n(\kappa', z' - z'') = \int_{-\infty}^{+\infty} dk_z \ \Phi_n(\kappa', \kappa_z) \cos[\kappa_z(z' - z'')], \tag{7.102}$$

---

[3] Similar calculations can be performed for $B_S(\rho)$.

where $\Phi(\kappa)$ is the three-dimensional spectrum for refractive index fluctuations. Inserting equation (7.101) into equation (7.100) and performing the $\kappa'$ integration, the covariance function $B_\chi(\rho)$ has the form:

$$B_\chi(\rho) = \int d^2\kappa\, e^{i\kappa\cdot\rho}\left\{ k^2 \int_0^z dz' \int_0^z dz'' \sin\left(\frac{\kappa^2(z-z')}{2k}\right)\sin\right.$$
$$\left.\left(\frac{\kappa^2(z-z'')}{2k}\right)F_n(K, z' - z'')\right\}. \tag{7.103}$$

Similarly, the covariance function $B_S(\rho)$ is given by:

$$B_S(\rho) = \int d^2\kappa\, e^{i\kappa\cdot\rho}\left\{ k^2 \int_0^z dz' \int_0^z dz'' \cos\left(\frac{\kappa^2(z-z')}{2k}\right)\cos\right.$$
$$\left.\left(\frac{\kappa^2(z-z'')}{2k}\right)F_n(\kappa, z' - z'')\right\}. \tag{7.104}$$

Therefore, the covariance function $B_\chi(\rho)$ and $B_S(\rho)$ are simply the Fourier transforms of the quantities inside the curly brackets in equations (7.103) and (7.104). These quantities represent the two-dimensional spectral densities of the amplitude and phase fluctuations $F_\chi(\kappa, 0)$ and $F_S(\kappa, 0)$ respectively:

$$F_\chi(\kappa, 0) = k^2 \int_0^z dz' \int_0^z dz'' \sin\left(\frac{\kappa^2(z-z')}{2k}\right)\sin\left(\frac{\kappa^2(z-z'')}{2k}\right)F_n(\kappa, z' - z''), \tag{7.105}$$

$$F_S(\kappa, 0) = k^2 \int_0^z dz' \int_0^z dz'' \cos\left(\frac{\kappa^2(z-z')}{2k}\right)\cos\left(\frac{\kappa^2(z-z'')}{2k}\right)F_n(\kappa, z' - z''). \tag{7.106}$$

Let us first try to reduce $F_\chi(\kappa, 0)$ and $F_S(\kappa, 0)$ to a simpler form. This can be achieved by making use of the properties of the $F_n(\kappa, z' - z'')$ as first shown by Tatarskii. It is clear from equation (7.102) that $F_n(\kappa, z' - z'')$ is an even function of $z' - z''$, that is

$$F_n(\kappa, z' - z'') = F_n(\kappa, z'' - z'), \tag{7.107}$$

which represents isotropic turbulence and so in equations (7.105) and (7.106) vector notation for $F_n$ can be dropped. Next we make a transformation of variables from $z'$ and $z''$ to the difference variable $\xi = z' - z''$ and center of mass variable $2\eta = z' + z''$ respectively. The region of integration for the new variables $(\xi, \eta)$ is a rhombus whose limits are given by the straight lines: $\eta = 0.5\xi$, $\eta = -0.5\xi$, $\eta = z + 0.5\xi$ and $\eta = z - 0.5\xi$ as shown in figure 7.6. Assuming the total propagation length $z = L$ and using the trigonometric identity for the product of two sine functions, the equation for $F_\chi(\kappa, 0)$ reduces to:

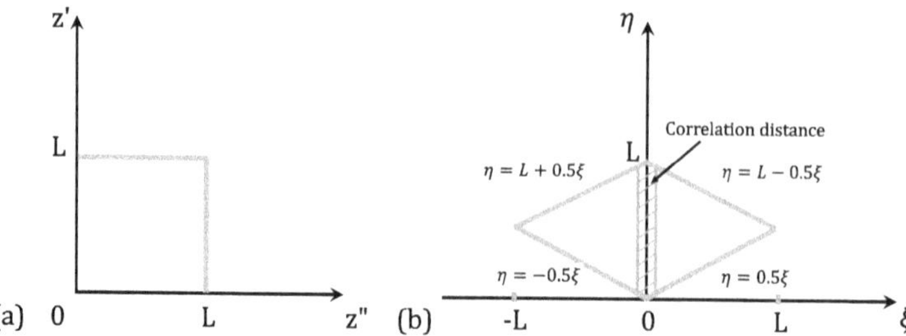



**Figure 7.6.** Range of integration shown for (a) $z'-z''$ coordinates and (b) $\eta-\xi$ coordinates. Note that the coordinate transformation from $z'-z''$ to $\eta-\xi$ changes the shape of the region of integration from a square to a rhombus. The equations for the sides of this rhombus are also given in (b). The shaded region shows the distance over which the correlation is significant.

$$F_\chi(\kappa, 0) = k^2 \int_0^L d\xi \, F_n(\kappa, \xi) \int_{\frac{\xi}{2}}^{L-\frac{\xi}{2}} d\eta \left[ \cos\left(\frac{\kappa^2 \xi}{2k}\right) - \cos\left(\frac{\kappa^2(L-\eta)}{k}\right) \right]. \quad (7.108)$$

As $F_n(\kappa, \xi)$ is a function of $\xi$ only, the $\eta$ integral can be evaluated independently. This gives us:

$$F_\chi(\kappa, 0) = \int_0^L d\xi \, F_n(\kappa, \xi) \left[ k^2(L - \xi)\cos\left(\frac{\kappa^2 \xi}{2k}\right) + \frac{k^3}{\kappa^2} \sin\left(\frac{\kappa^2(\xi)}{2k}\right) \right.$$
$$\left. - \frac{k^3}{\kappa^2} \sin\left(\frac{\kappa^2(2L - \xi)}{2k}\right) \right]. \quad (7.109)$$

The two-dimensional spectral density $F_n(\kappa, \xi)$ represents the correlation of the refractive index fluctuations $n_1$ on any two adjacent planes $z'$ and $z''$. If the separation between these two places is large, then smaller eddies would not intersect both planes and so will contribute negligibly to the correlation. Therefore, only contribution from those turbulent eddies is significant whose spatial size, $l = 2\pi\kappa^{-1}$ is less than or equal to the separation $\xi$. The implies that the function $F_n(\kappa, z' - z'')$ falls quickly to zero for $\kappa\xi \geqslant 1$. The maximum value of $\kappa$ ($\kappa_{\mathrm{max}}$) is given by the smallest eddy size or the lower scale of the turbulence $l_0$. Therefore, the important range of integration is given by $\xi \leqslant \kappa^{-1} \leqslant \kappa_{\mathrm{max}}^{-1}$. We had earlier assumed that the wavelength of the beam is much less than the $l_0$, implying that the quantity $\kappa^2\xi/k \leqslant \lambda/l_0 \ll 1$. Apart from these conditions, we also have that the integral above has major contributions only for values of $\xi \ll L$. Since we are interested in correlation of $\chi$ and $S$ over such small distances, we can approximate the trigonometric quantities in equation (7.109) as $\cos(\kappa^2\xi/2k) \approx 1$, $\sin(\kappa^2\xi/2k) \approx \kappa^2\xi/2k$ and $\sin(\kappa^2(2L - \xi)/2k) \approx \sin(\kappa^2L/k)$, and write:

$$F_\chi(\kappa, 0) = \left[ k^2L - \frac{k^3}{\kappa^2} \sin\left(\frac{\kappa^2 L}{k}\right) \right] \int_0^L F_n(\kappa, \xi) \, d\xi. \quad (7.110)$$

The function $F_n(\kappa, \xi)$ has an appreciable value only over the correlation distance (see figure 7.6), that is when $\xi \ll L$. So the integration limits in equation (7.110) can be changed to $+\infty$ without much change in the value of the $\xi$ integral. Thus, the final form of the two-dimensional spectral density function for the amplitude fluctuations may be written as:

$$F_\chi(\kappa, 0) = \pi k^2 L \left[ 1 - \frac{k}{\kappa^2 L} \sin\left(\frac{\kappa^2 L}{k}\right) \right] \Phi_n(\kappa). \qquad (7.111)$$

Similarly, the two-dimensional spectral density function for the phase fluctuations is given by:

$$F_S(\kappa, 0) = \pi k^2 L \left[ 1 + \frac{k}{\kappa^2 L} \sin\left(\frac{\kappa^2 L}{k}\right) \right] \Phi_n(\kappa). \qquad (7.112)$$

These expressions for the two-dimensional spectral density of the amplitude and phase fluctuations can be substituted in equations (7.103) and (7.104) to evaluate the covariance functions $B_\chi(\rho)$ and $B_S(\rho)$ respectively. However, before that it is important to understand the various scales associated with $F_\chi(\kappa, 0)$ and $F_S(\rho)$. Consider the two-dimensional spectral density for the amplitude fluctuations $F_\chi(\kappa, 0)$ which is a product of two functions:

(a) The three-dimensional spectrum of the refractive index fluctuations $\Phi_n(\kappa)$, and

(b) The function $f(\kappa) = \left[ 1 - \frac{k}{\kappa^2 L} \sin\left(\frac{\kappa^2 L}{k}\right) \right].$

The function $\Phi_n(\kappa)$ is defined for the spatial frequencies lying in the range bounded by the inner and outer scales of turbulence: $\frac{2\pi}{L_o} < \kappa < \frac{2\pi}{l_o}$. For spatial scales which are much smaller than the inner scale $l_o$ (or $\kappa \gg \frac{2\pi}{l_o}$), the value of $\Phi_n(\kappa)$ is negligibly small. Similarly, the function $\Phi_n(\kappa)$ loses its meaning for spatial scales larger than the outer scale of turbulence $L_o$ as for such large eddies the fundamental assumption of an isotropic and homogeneous turbulence is no longer true. Similarly, consider function $f(\kappa)$ for $\frac{\kappa^2 L}{k} \ll 1$ where the sine function can be expanded into a Taylor series:

$$f(\kappa) = 1 - \frac{k}{\kappa^2 L} \left[ \left(\frac{\kappa^2 L}{k}\right) - \frac{1}{3!}\left(\frac{\kappa^2 L}{k}\right)^3 + \cdots \right] \approx \frac{1}{6}\frac{\kappa^4 L^2}{k^2}. \qquad (7.113)$$

On the other hand, for large values of $\kappa$, the sine function oscillates fast, in turn contributing negligibly to $f(\kappa)$ and thus the function $f(\kappa) \to 1$. A characteristic scale $\kappa_o = \frac{2\pi}{\sqrt{\lambda L}}$ can be defined at which $\sin\left(\frac{\kappa_o^2 L}{k}\right) = 0$ and $f(\kappa) = 1$. Therefore, we can write the function $f(\kappa)$ as:

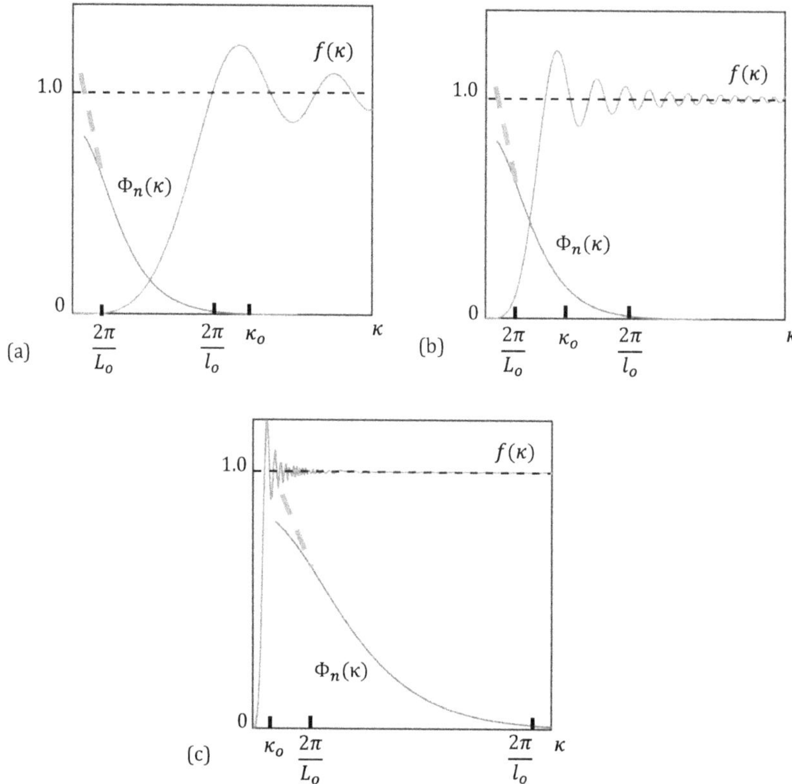

**Figure 7.7.** The variation of functions $\Phi_n(\kappa)$ and $f(\kappa)$ as a function of the spatial frequency $\kappa$ are shown for different cases of $\kappa_o$, $l_o$ and $L_o$. The three different cases: (a) $l_o > \sqrt{\lambda L}$; (b) $L_o > \sqrt{\lambda L} > l_o$ and (c) $L_o < \sqrt{\lambda L}$ are shown here.

$$f(\kappa) = \begin{cases} \dfrac{1}{6}\dfrac{\kappa^4 L^2}{k^2}, & \kappa < \kappa_o \\ 1, & \kappa > \kappa_o. \end{cases} \tag{7.114}$$

Depending on the value of the Fresnel length $\sqrt{\lambda L}$ with respect to the $l_o$ and $L_o$, there are three different cases for the value of two-dimensional spectral density $F_\chi(\kappa, 0)$ that can be studied:

(a) $l_o > \sqrt{\lambda L}$

  Figure 7.7(a) shows this particular case. It can be seen that the quantity $\kappa_o$ is greater than the maximum allowed spatial frequency $\kappa_{\text{max}} \sim \dfrac{2\pi}{l_o}$ for the refractive index spectrum $\Phi_n(\kappa)$. The spectrum $\Phi_n(\kappa) \to 0$ in the region where $\kappa > \kappa_{\text{max}}$. From the properties of the function $f(\kappa)$, it can be approximated by $\dfrac{1}{6}\dfrac{\kappa^4 L^2}{k^2}$ for all values of $\kappa < \kappa_o$. Thus, we can write the two-dimensional spectral density of the amplitude fluctuations $F_\chi(\kappa, 0)$ as:

$$F_\chi(\kappa, 0) = \begin{cases} \dfrac{1}{6}\pi L^3 \kappa^4 \Phi_n(\kappa), & \kappa < \kappa_{\max} \\ 0, & \kappa > \kappa_{\max}. \end{cases} \tag{7.115}$$

The product of functions $\Phi_n(\kappa)$ and $f(\kappa)$ will have a maximum value near $\kappa_{\max}$. This means that for the case when $l_o > \sqrt{\lambda L}$, the refractive index inhomogeneities with sizes of the order of $l_o$ will have the greatest effect on the amplitude fluctuations. The covariance function $B_\chi(\rho)$ in this case would therefore have a correlation length of the order of $l_o$.

(b) $L_o > \sqrt{\lambda L} > l_o$

In this case, the product of the functions $\Phi_n(\kappa)$ and $f(\kappa)$ has a maximum near the point $\kappa_o$ and becomes zero for $\kappa > \kappa_{\max}$ as can clearly be seen in figure 7.7(b). Therefore, $F_\chi(\kappa, 0)$ can be written as:

$$F_\chi(\kappa, 0) = \begin{cases} \dfrac{1}{6}\pi L^3 K^4 \Phi_n(\kappa), & \kappa < \kappa_o \\ \pi k^2 L \Phi_n(\kappa), & \kappa_o < \kappa < \kappa_{\max} \\ 0, & \kappa > \kappa_{\max}. \end{cases} \tag{7.116}$$

In this case, refractive index inhomogeneities of the spatial size $\sqrt{\lambda L}$ cause the major amplitude fluctuations in the light beam. At the final detector plane situated at $x = L$, the correlation function $B_\chi(\rho)$ would show a correlation distance of the order of $\sqrt{\lambda L}$ or the Fresnel length for the particular propagation geometry. Notice that in figures 7.7(a) and (b), the function $f(\kappa)$ is near zero for all frequencies $\kappa < \kappa_{\min}$ where $\kappa_{\min} \sim \dfrac{2\pi}{L_o}$.

(c) $L_o < \sqrt{\lambda L}$

In this case, the major influence on the amplitude fluctuations comes from refractive index inhomogeneities with spatial sizes between $L_o$ and $\sqrt{\lambda L}$. However, the refractive index eddies whose sizes are greater than the outer scale $L_o$ cannot be considered isotropic or homogeneous. The correlation function between the refractive index inhomogeneities is now no longer independent of the location of the observation points. Therefore, now the spectral density $\Phi_n(\kappa)$ is no longer defined in this region. The $\Phi_n(\kappa)$ has meaning only in the region when $\kappa \geqslant \kappa_{\min}$. So we can write the function $F_\chi(\kappa, 0)$ as:

$$F_\chi(\kappa, 0) = \pi k^2 L \Phi_n(\kappa), \quad \kappa > \kappa_{\min}. \tag{7.117}$$

In the case of $F_S(\kappa, 0)$, the function $f(\kappa) = \left[1 + \dfrac{k}{\kappa^2 L}\sin\left(\dfrac{\kappa^2 L}{k}\right)\right]$ has a maximum at $\kappa = 0$. This means that the refractive index eddies of scale sizes $L_o$ greatly influence the phase fluctuations which means that in order to get stable statistics, long measurement times are required. However, these eddies are non-stationary in space as the atmosphere is dynamically changing at millisecond time scales, thus corrupting the measured data. Thus, it is not practical to measure the covariance function $B_S(\rho)$. Instead, the phase structure function $D_S(\rho)$ defined as:

$$D_S(\rho) = 2[B_S(0) - B_S(\rho)],  \tag{7.118}$$

offers a more physically useful measure of the phase fluctuations. The covariance functions $B_\chi(\rho)$ and $B_S(\rho)$ can be obtained by taking the Fourier transform of its respective spectral density functions $F_\chi(\kappa, 0)$ and $F_S(\kappa, 0)$. We can substitute $F_\chi(\kappa, 0)$ and $F_S(\kappa, 0)$ from equations (7.111) and (7.112) into equations (7.103) and (7.104) to obtain:

$$B_{\chi,S}(\rho) = \pi k^2 L \int d^2\kappa \, e^{i\kappa\cdot\rho}\left[1 \mp \frac{k}{\kappa^2 L}\sin\left(\frac{\kappa^2 L}{k}\right)\right]\Phi_n(\kappa).  \tag{7.119}$$

We can represent $\kappa$ in polar coordinates and substitute $\kappa = (\kappa, \theta)$ and $d^2\kappa = \kappa\, dk\, d\theta$ in the above equation. Further, as we are dealing with isotropic and homogeneous turbulence, we use the Bessel identity (see equation (7.94)) to evaluate the $\theta$ integral and get:

$$B_{\chi,S}(\rho) = 2\pi^2 k^2 L \int_0^\infty \kappa J_o(\kappa\rho)\left[1 \mp \frac{k}{\kappa^2 L}\sin\left(\frac{\kappa^2 L}{k}\right)\right]\Phi_n(\kappa)d\kappa.  \tag{7.120}$$

The phase structure function $D_s(\rho)$ can thus be given as:

$$D_S(\rho) = 4\pi^2 k^2 L \int_0^\infty \kappa\left[1 + \frac{k}{\kappa^2 L}\sin\left(\frac{\kappa^2 L}{k}\right)\right][1 - J_o(\kappa\rho)]\Phi_n(\kappa)d\kappa.  \tag{7.121}$$

Similar calculations for the log-amplitude structure function $D_\chi(\rho)$ yields the relation,

$$D_\chi(\rho) = 4\pi^2 k^2 L \int_0^\infty \kappa\left[1 - \frac{k}{\kappa^2 L}\sin\left(\frac{\kappa^2 L}{k}\right)\right][1 - J_o(\kappa\rho)]\Phi_n(\kappa)d\kappa.  \tag{7.122}$$

The total wave structure function is equal to the sum of the log-amplitude and phase structure functions,

$$D(\rho) = D_\chi(\rho) + D_S(\rho),  \tag{7.123}$$

which for the case of unbounded plane waves is obtained from equations (7.121) and (7.122) as,

$$D(\rho) = 8\pi^2 k^2 L \int_0^\infty \kappa\,\Phi_n(\kappa)[1 - J_o(\kappa\rho)]d\kappa.  \tag{7.124}$$

The particular form for the amplitude covariance function and the phase structure function is obtained by plugging the desired refractive index spectrum into the above equations. Substituting Kolmogorov spectrum for $\Phi_n(\kappa)$, we get:

$$B_\chi(\rho) = 2\pi^2(0.033)C_n^2 k^2 L \int_0^\infty d\kappa \; J_o(\kappa\rho)\left[1 - \frac{k}{\kappa^2 L}\sin\left(\frac{\kappa^2 L}{k}\right)\right]\kappa^{-8/3}, \qquad (7.125)$$

$$D_S(\rho) = 4\pi^2(0.033)C_n^2 k^2 L \int_0^\infty d\kappa\left[1 + \frac{k}{\kappa^2 L}\sin\left(\frac{\kappa^2 L}{k}\right)\right][1 - J_o(\kappa\rho)]\kappa^{-8/3}. \qquad (7.126)$$

Assuming that $l_o \gg \sqrt{\lambda L}$, we can calculate the variance of the amplitude fluctuations $\sigma_\chi^2$ from the covariance function as:

$$\sigma_\chi^2 = B_\chi(\rho = 0) = 0.31 C_n^2 k^{7/6} L^{11/6}. \qquad (7.127)$$

The power law for the phase structure function for the different conditions is then obtained as:

$$D_s(\rho) = \begin{cases} 1.46 C_n^2 k^2 L\rho^{5/3}, & \text{when } l_o \ll \rho \ll \sqrt{\lambda L}, \\ 2.91 C_n^2 k^2 L\rho^{5/3}, & \text{when } \sqrt{\lambda L} \ll \rho \ll L_o. \end{cases} \qquad (7.128)$$

*Mutual coherence function*

The mutual coherence function or the second moment for the field at some propagation distance $L$ is given by the ensemble average,

$$\Gamma_2(r_1, r_2, L) = \langle U(r_1, L) \; U^*(r_2, L)\rangle, \qquad (7.129)$$

where $U^*(r, L)$ denotes the complex conjugate field. We substitute the field $U(r, L)$ from equation (7.51) to get,

$$\Gamma_2(r_1, r_2, L) = U_0(r_1, L) \; U_0^*(r_2, L) \; \langle\exp[\psi(r_1) + \psi^*(r_2)]\rangle \qquad (7.130)$$

The mutual coherence function, when evaluated at identical observation points $r_1 = r_2 = r$ gives the mean irradiance value,

$$\langle I(r, L)\rangle = \Gamma_2(r, r, L). \qquad (7.131)$$

The mean irradiance can be used to predict the atmosphere-induced beam spreading [3]. For the case of an infinite plane wave with no wavefront curvature, the mutual coherence function is given by the expression,

$$\Gamma_2(\rho, L) = \exp\left[-4\pi^2 k^2 L \int_0^\infty \kappa \; \Phi_n(\kappa)[1 - J_o(\kappa\rho)]d\kappa\right], \qquad (7.132)$$

where $\rho = |r_1 - r_2|$ is the separation distance between two points on the wavefront. The mutual coherence function can also be used to predict the spatial coherence radius at the detector plane. The modulus of the complex degree of coherence

(DOC), which provides information regarding the loss of spatial coherence of an initially coherent beam is given by,

$$\text{DOC}(r_1, r_2, L) = \frac{|\Gamma_2(r_1, r_2, L)|}{\sqrt{\Gamma_2(r_1, r_1, L)\, \Gamma_2(r_1, r_1, L)}}, \tag{7.133}$$

$$= \exp\left[-\frac{1}{2}D(r_1 r_2, L)\right]. \tag{7.134}$$

Here, $D(r_1 r_2, L)$ is the wave structure function given by equation (7.124). The spatial coherence radius $\rho_o$ (measured in meters) is defined by the separation distance at which the DOC reduces to value $1/e$, i.e. when $D(\rho, L) = 2$. For the unbounded plane wave, we get

$$\rho_o = \begin{cases} \left[1.46k^2 \int_0^L C_n^2(z)dz\right]^{-3/5}, & \text{when } l_o \ll \rho_o \ll L_o, \\ [1.46C_n^2 k^2 L]^{5/3}, & \text{when } l_o \ll \rho_o \ll L_o \;\; C_n^2 = \text{Constant}. \end{cases} \tag{7.135}$$

Sometimes, the spatial coherence is defined through the atmospheric coherence width $(r_o)$ which is equal to $r_o = 2.1\rho_o$. This term, also commonly known as the Fried parameter, was first introduced by Fried [26, 27] who showed that it is an important measure of the performance of an imaging system. In the presence of turbulence, the angular extent (seeing angle) of the far-field diffraction pattern is given as $\lambda/r_o$. The Fried parameter can be interpreted as the aperture size beyond which further increases in diameter result in no further increase in resolution. The Fried parameter for plane waves is given as,

$$r_{op} = \left[0.42k^2 \int_0^L C_n^2(z)dz\right]^{-3/5}. \tag{7.136}$$

*Cross-coherence function*

The fluctuations in the irradiance of the field are described by the cross-coherence function or the fourth-order moment of the field,

$$\Gamma_4(r_1, r_2, r_3, r_4, L) = \langle U(r_1, L)U^*(r_2, L)U(r_3, L)U^*(r_4, L)\rangle, \tag{7.137}$$

which can be expressed in the form,

$$\Gamma_4(r_1, r_2, r_3, r_4, L) = U_0(r_1, L)U_0^*(r_2, L)U_0(r_3, L)U_0^*(r_4, L) \\ \times \langle \exp[\psi(r_1) + \psi^*(r_2) + \psi(r_3) + \psi^*(r_4)]\rangle. \tag{7.138}$$

By setting $r_1 = r_2 = r_3 = r_4 = r$, the fourth-order coherence function yields the second moment of irradiance,

$$\langle I^2(r, L)\rangle = \Gamma_4(r, r, r, r, L). \tag{7.139}$$

The covariance function of irradiance is defined by the normalized quantity,

$$B_I(r_1, r_2, L) = \frac{\Gamma_4(r_1, r_1, r_2, r_2, L) - \Gamma_2(r_1, r_1, L)\Gamma_2(r_2, r_2, L)}{\Gamma_2(r_1, r_1, L)\Gamma_2(r_2, r_2, L)}, \tag{7.140}$$

$$= \frac{\Gamma_4(r_1, r_1, r_2, r_2, L)}{\Gamma_2(r_1, r_1, L)\Gamma_2(r_2, r_2, L)} - 1, \tag{7.141}$$

which for the case $r_1 = r_2 = r$ gives us the scintillation index,

$$\sigma_I^2(r, L) = \frac{\langle I^2(r, L)\rangle}{\langle I(r, L)\rangle} - 1. \tag{7.142}$$

Scintillation refers to the temporal and spatial fluctuations in the irradiance of a beam on propagation through atmospheric turbulence. When the log-amplitude variance $\sigma_\chi^2$ is sufficiently small, i.e. $\sigma_\chi^2 \ll 1$, then

$$\sigma_I^2(r, L) = \exp\left[4\sigma_\chi^2(r, L)\right] - 1, \tag{7.143}$$

$$\cong 4\sigma_\chi^2(r, L). \tag{7.144}$$

The scintillation index can be expressed as a sum of radial and longitudinal components,

$$\sigma_I^2(r, L) = \sigma_{I,\mathrm{rad}}^2(r, L) + \sigma_{I,\mathrm{long}}^2(r, L). \tag{7.145}$$

The radial component of the scintillation index, $\sigma_{I,\mathrm{rad}}^2(r, L)$ is zero at the beam center $r = 0$. This component is also zero for the case of an infinite plane or spherical wave. The longitudinal component of the scintillation index $\sigma_{I,\mathrm{long}}^2(r, L)$ is constant across the beam cross section in any $z$ plane. The longitudinal component corresponds to the on-axis scintillation index and is also written as $\sigma_I^2(0, L)$. Using equation (7.127) as the form of $\sigma_\chi^2$, the scintillation index for a plane wave is given as,

$$\sigma_R^2 = 1.23 C_n^2 k^{7/6} L^{11/6}. \tag{7.146}$$

This quantity is also commonly known as Rytov variance [3]. The Rytov variance value can be used as a theoretical measure to distinguish between weak or strong refractive fluctuation regimes as shown in table 7.1.

*Validity of Rytov method*
The assumption that $|\nabla\psi_1| \ll |\nabla\psi_0|$ leads to important restrictions on the range of validity of the Rytov method. When the Rytov method was first developed, it appeared to give results which were in good agreement to the experimental data, that were measured over atmospheric propagation paths of less than 1 km. This lead to the belief that this method gave amplitude and phase results for a much greater range of validity in terms of $C_n^2$ and propagation path compared to the Born approximation. Later, studies by several researchers [28–31] challenged this notion

**Table 7.1.** Classification of refractive index fluctuations on the basis of Rytov variance.

| Fluctuation condition | Rytov variance value |
| --- | --- |
| Weak | $\sigma_I^2 < 1$ |
| Moderate | $\sigma_I^2 \sim 1$ |
| Strong | $\sigma_I^2 > 1$ |
| Saturation | $\sigma_I^2 \rightarrow \infty$ |

of the extended range of validity of this method. It is now well established that the results obtained through the Rytov approach are valid only when $\sigma_R^2 \leqslant 0.3$ [1, 32]. Therefore, it can be applied to very short distances when the turbulence is strong. However, at night, when the value of $C_n^2$ may drop up to two orders of magnitude, this method may sometimes provide good estimates even for propagation paths of 100 km. The method would also be applicable for longer ranges when the light wavelength is an order of magnitude or more than the optical range. Also, some experimental work suggests that in strong integrated turbulence, the phase statistics can have a larger range of validity than log-amplitude statistics [18, 32, 33]. Newer methods like the parabolic equation method, the diagram method, the coherence theory based approach and the Markov approximation have also been explored in this regard [5]. Recently, the extended Huygens–Fresnel (eHF) method [34–36] has also emerged as a popular approach for studying the propagation of a beam through random media like atmospheric turbulence. We will describe the eHF approach in some detail in the next section.

## 7.4 Extended Huygens–Fresnel integral approach

According to the Huygens construction, in the propagation of any beam through vacuum, every point on the wavefront can be considered as a center of a secondary disturbance which gives rise to secondary wavelets. At any later instant, the wavefront may be regarded as the envelope of these wavelets. Fresnel supplemented Huygens' principle by providing an explanation for the process of diffraction by stating that these secondary wavelets mutually interfere. The combination of Huygens' construction with the principle of interference is together called the *Huygens–Fresnel principle*. It was shown independently by Feizulin and Kravtsov [35] in 1967 and by Lutomirski and Yura [34] in 1971 that one can develop an extension of the Huygens–Fresnel principle for propagation in a weakly inhomogeneous medium. They showed that for a refractive medium, the secondary wavefront will again be determined by the envelope of spherical wavelets originating from the primary wavefront. However, now each wavelet will be determined by its propagation through the inhomogeneous medium. Starting from the stochastic Helmholtz equation, Lutomirski and Yura followed a method very similar to the integral theorem of Helmholtz and Kirchhoff used in free space diffraction theory and showed that the field at any arbitrary point $P_0$ inside the medium can be expressed in terms of the boundary values of the wave on any closed surface surrounding that point. Let the turbulent medium occupy a volume $V$ which is bounded by a surface $S$ and the refractive index variation inside the volume is given by equation (7.6). Then, within the volume $V$, let a point source be situated at $P_0$ which produces a field $G(P_1, P_0)$ at any point $P_1$, then using the equation (7.8) one obtains:

$$\left[\nabla^2 + k^2 n^2(r)\right] G(P_1, P_0) = -4\pi\delta(|P_1 - P_0|), \tag{7.147}$$

where $\delta$ is the Dirac Delta function. Multiply the equation (7.8) by $G$ and the equation (7.147) by $U$ and subtract them to yield,

$$G(P_1, P_0)\nabla^2 U - U\nabla^2 G(P_1, P_0) = 4\pi\delta(|r - P|)U. \tag{7.148}$$

If the field $U$ possesses continuous first- and second-order derivatives inside and on $S$, then one can integrate equation (7.148) over $V$ and apply the Green theorem,

$$U(P_0) = \frac{1}{4\pi} \iint_S (U\nabla G - G\nabla U) \cdot dA. \tag{7.149}$$

Here, $dA$ represents a surface element with its normal directed into $V$. We want to find the complex field distribution at some arbitrary point in a turbulent medium when the optical field specified over a finite surface like the aperture, is available. Therefore, let a monochromatic optical disturbance pass through an aperture $A$ in an opaque screen and propagate into the turbulent medium. The aperture is assumed to be large compared to the wavelength of light but smaller than the distance between the observation point $P_0$ and the opaque screen. Now again following arguments analogous to the free space Kirchhoff formulation of diffraction through a planar screen, equation (7.149) is reduced into an integral over an aperture of area $\Sigma$,

$$U(P_0) = \frac{1}{4\pi} \iint_\Sigma (U\nabla G - G\nabla U) \cdot dA. \tag{7.150}$$

Let the optical disturbance propagate along the $z$-axis and the aperture $A$ lies in a plane normal to this direction. The unit vector normal to the beam wavefront in the aperture is denoted by $\hat{n}$, then $\partial U/\partial z \approx ikU(\hat{n} \cdot \hat{z})$. Next, the function $G$ is chosen to be of the form $G = \exp(ikr_{01} + \psi)/r_{01}$ where $r_{01}$ is the distance between the point $P_0$ and the elemental area $dA$ in the aperture, given by $s = |P_0 - P_1|$. Following Kirchhoff treatment and using the assumption that $s \gg \lambda$, one can write,

$$U(P_0) = \frac{ik}{4\pi} \iint_\Sigma U(P_1)\frac{\exp(ikr_{01} + \psi)}{r_{01}}\left[\hat{r}_{01} \cdot \hat{z} - \hat{n} \cdot \hat{z} + \frac{\nabla\psi}{ik} \cdot \hat{z}\right] d^2r, \tag{7.151}$$

where $\hat{r}_{01} = (P_0 - P_1)/|P_0 - P_1|$ and $d^2r$ is an element of area at the point $P_1$ in the aperture. In the geometric optics regime, one can write $\psi \sim ik \int n_1 ds$, then $|(\nabla\psi/ik) \cdot \hat{z}| = |n_1(\hat{s} \cdot \hat{z})|$ where $n_1 \sim 10^{-6}$. The third term in the integral is neglected as it is very small compared to unity for all distances of interest [34]. The area $\Sigma$ can be replaced with a portion of the incident wavefront which fills the aperture.

Over this new region of integration $A$, the derivative of the field $U$ in the direction normal to the surface is $\partial U/\partial z \leqslant ikU$. This implies that the equation (7.150) can be written as,

$$U(P_0) = \int_A K(h)G(P'_1, P_0)U(r')d^2r', \tag{7.152}$$

where $K(h) = (-i/2\lambda)(1 + \cos(h))$ and $h = \pi - \cos^{-1}(\hat{n} \cdot \hat{z})$. Remember that $G(P_1, P_0)$ was defined as the field at point $P_1$ due to a point source situated at $P_0$. However, it was shown by Yura that the field and source points can be interchanged as $G(P_1, P_0) = G(P_0, P_1)$ [34]. This gives the extended form of the Huygens–Fresnel principle for an inhomogeneous medium where the element of area $dA$ of the wavefront contributes to the field at point $P_0$. Therefore, if one knows how a spherical wave propagates in a given medium, the response to a random disturbance in the aperture can be determined.

The most commonly used form for the extended Huygens–Fresnel principle is obtained by further assuming paraxially propagating waves and using the Fresnel diffraction as the free space propagator. In that case, the extended Huygens–Fresnel principle is given by:

$$U(r_1) = -\frac{ik}{2\pi L} \exp(ikL) \iint_{-\infty}^{+\infty} U(r_0) \exp\left[\left(\frac{ik \, | \, r_1 - r_0 \, |^2}{2L}\right) + \psi(r, r_0)\right] d^2r_0. \tag{7.153}$$

However, it was recently shown by Charnotskii [36] that the extended Huygens–Fresnel principle is actually a rigorous consequence of the Helmholtz wave equation and it requires no additional assumptions like (a) the paraxial approximation or (b) weak turbulence. In this case, the kernel of the operator is not the Green function of the original Helmholtz equation but rather the doubled derivative of the Green function along the propagation direction.

In equation (7.153), the randomness of the medium is contained in the complex phase term $\psi(r_1, r_0)$ which is being added to each spherical wave. This is a very general description of the turbulent effects on the propagation. The usual practice is to use Rytov approximation to describe the phase term $\psi(r_1, r_0)$,

$$\exp(\psi) = \exp(\psi_1 + \psi_2), \tag{7.154}$$

where $\psi_1$ and $\psi_2$ are the first- and second-order Rytov approximation terms This method is unreliable when the complex phase disturbances $\langle \psi \rangle^2 > 1$. Charnotskii has described the various regions of validity of the extended Huygens–Fresnel principle [36]. He has suggested that when the phase term $\psi$ is represented through equation (7.154), then the resultant field violates the energy-conservation principle and in that case the accuracy of the scintillation calculations is uncertain. This form of $\psi$ fetches accurate results for the fourth moment only under weak scintillation conditions.

# References


[1] Strohbehn J W 1968 Line-of-sight wave propagation through the turbulent atmosphere *Proc. IEEE* **56** 1301–8

[2] Tatarskii V I 1961 *Wave Propagation in a Turbulent Medium* (New York: McGraw-Hill) (volume translated by Richard A Silverman)

[3] Andrews L C and Phillips R L 2005 *Laser Beam Propagation Through Random Media* (Bellingham, WA: SPIE Optical Engineering Press)

[4] Wheelon A D 2001 *Electromagnetic Scintillation* vol 1 (Cambridge: Cambridge University Press)

[5] Strohbehn J W 1978 *Laser Beam Propagation in the Atmosphere (Topics in Applied Physics* vol 25) (Berlin: Springer)

[6] Mandel L and Wolf E 1995 *Optical Coherence and Quantum Optics* (Cambridge: Cambridge University Press)

[7] Roggemann M C and Welsh B M 1996 *Imaging Through Turbulence* (Boca Raton, FL: CRC Press)

[8] Goodman J W 2015 *Statistical Optics* 2nd edn (New York: Wiley)

[9] Champagne F H, Friehe C A, LaRue J C and Wynagaard J C 1977 Flux measurements, flux estimation techniques, and fine-scale turbulence measurements in the unstable surface layer over land *J. Atmos. Sci.* **34** 515–30

[10] Williams R M and Paulson C A 1977 Microscale temperature and velocity spectra in the atmospheric boundary layer *J. Fluid Mech.* **83** 547–67

[11] Oboukhov A M 1949 Structure of the temperature field in turbulent flow *Izv. Akad. Nauk. SSSR, Ser. Geogr. i Geofiz.* **13** 58–69

[12] Corrsin S 1951 On the spectrum of isotropic temperature fluctuations in an isotropic turbulence *J. Appl. Phys.* **22** 469–73

[13] Hill R J 1978 Models of the scalar spectrum for turbulent advection *J. Fluid Mech.* **88** 541–62

[14] Hill R J and Clifford S F 1978 Modified spectrum of atmospheric temperature fluctuations and its application to optical propagation *J. Opt. Soc. Am.* **68** 892–9

[15] Churnside J H and Lataitis R J 1990 Wander of an optical beam in the turbulent atmosphere *Appl. Opt.* **29** 926–30

[16] Andrews L C 1992 An analytical model for the refractive index power spectrum and its application to optical scintillations in the atmosphere *J. Mod. Opt.* **39** 1849–53

[17] Tatarskii V I 1967 Depolarization of light by turbulent atmospheric inhomogeneities *Radiophys. Quantum Electron.* **10** 987–8

[18] Strohbehn J and Clifford S 1967 Polarization and angle-of-arrival fluctuations for a plane wave propagated through a turbulent medium *IEEE Trans. Antennas Propag.* **15** 416–21

[19] Saleh A A M 1967 An investigation of laser wave depolarization due to atmospheric transmission *IEEE J. Quantum Electron.* **3** 540–3

[20] Cox D C, Arnold H W and Hoffman H H 1981 Observations of cloud-produced amplitude scintillation on 19- and 28-GHz earth-space paths *Radio Sci.* **16** 885–907

[21] Rytov S M 1937 Diffraction of light by ultrasonic waves *Izv. Akad. Nauk. SSSR, Ser. Fiz. (Bull. Acad. Sci. USSR, Phys. Ser.)* **2** 223–59

[22] Obukhov A M 1953 Effect of weak inhomogeneities in the atmosphere on sound and light propagation *Izv. Akad. Nauk SSSR Ser. Geophys.* **2** 155–65

[23] Tatarskii V I 1971 *The Effects of the Turbulent Atmosphere on Wave Propagation* (Washington, DC: National Oceanic and Atmospheric Administration, U.S. Department of Commerce and the National Science Foundation)

[24] Yura H T, Sung C C, Clifford S F and Hill R J 1983 Second-order Rytov approximation *J. Opt. Soc. Am.* **73** 500–2

[25] Yura H T 1969 Optical propagation through a turbulent medium *J. Opt. Soc. Am.* **59** 111–2

[26] Fried D L 1966 Optical resolution through a randomly inhomogeneous medium for very long and very short exposures *J. Opt. Soc. Am.* **56** 1372–9

[27] Fried D L 1967 Optical heterodyne detection of an atmospherically distorted signal wave front *Proc. IEEE* **55** 57–77

[28] Brown W P 1967 Validity of the Rytov approximation *J. Opt. Soc. Am.* **57** 1539–42

[29] Lee R W and Harp J C 1969 Weak scattering in random media, with applications to remote probing *Proc. IEEE* **57** 375–406

[30] Tatarskii V I 1962 Second approximation in the problem of wave propagation in random medium *Izv. VUZ- Radiofiz.* **5** 490–507

[31] Hufnagel R E and Stanley N R 1964 Modulation transfer function associated with image transmission through turbulent media *J. Opt. Soc. Am.* **54** 52–61

[32] Fante R L 1975 Electromagnetic beam propagation in turbulent media *Proc. IEEE* **63** 1669–92

[33] Kallistratova M A, Gurvich A S and Time N S 1968 Fluctuations in the parameters of a light wave from a laser during propagation in the atmosphere *Radiophys. Quantum Electron.* **11** 771–6

[34] Lutomirski R F and Yura H T 1971 Propagation of a finite optical beam in an inhomogeneous medium *Appl. Opt.* **10** 1652–8

[35] Feizulin Z I and Kravtsov Y A 1967 Broadening of a laser beam in a turbulent medium *Radiophys. Quantum Electron.* **10** 33–5

[36] Charnotskii M 2015 Extended Huygen's Fresnel principle and optical waves propagation in turbulence: discussion *J. Opt. Soc. Am.* A **32** 1357–65

# Chapter 8

# Numerical simulation of laser beam propagation through turbulence

In the last chapter, we provided a detailed discussion of the various theoretical methodologies that have been developed for solving the stochastic Helmholtz equation [1]. These analytical methods were based on the approximation that the wavelength of the optical signal is significantly shorter than the typical size of the turbulent eddies in the atmosphere [2]. So the wave undergoes only forward scattering at small angles with respect to the propagation axis. The earlier perturbation series methods like Born's and Rytov's [3–8] expressed the statistics of the field $U$ in terms of the statistics of the fluctuating part $n_1$ of the refractive index. However, these methods are valid as long as the intensity fluctuations in the beam are smaller than $\sim 0.1$ times the mean intensity value [9]. This severely limited their usability to small propagation distances and low to moderate turbulence strengths. To extend the range of the validity of the wave-field solutions into the strong intensity fluctuation regimes, newer approaches like the diagram method, Markov method, and the local method of small approximations [7] were developed. These methods were useful in providing relations for the first and second moments of the field; however, the equations describing higher-order moments remained intractable with limited asymptotic solutions for the fourth moment of the field [6, 10, 11]. Recently, newer asymptotic schemes for determining intensity statistics for waves in turbulence with pure power-law spectra have been proposed [12, 13]. However, a major issue with these asymptotic schemes is that they assume unrealistically strong levels of turbulence. Another popular approach is the use of the extended Huygen's Fresnel principle for understanding the wave propagation phenomena and developing a heuristic theory for propagation through atmosphere [14–16]. This method has been able to provide many useful insights, but the results obtained are mainly qualitative in nature, and the accuracy of the higher-order moments is not yet ascertained. Therefore, between the weak turbulence regime where perturbation methods work and unrealistically strong turbulence regime where asymptotic solutions are provided, an entire range of realistic moderate to strong turbulence regions exists, which is not entirely understood by the current wave propagation theories. In the absence of a satisfactory theoretical description for the intensity fluctuations in this region, numerical simulations offer an attractive alternative approach that has proved to be valuable in practice. Figure 8.1 provides a comparison of the range of validity of various methods for laser propagation in turbulence.

Numerical simulation studies offer various advantages over the analytical methods for solving the stochastic wave equation. When applicable, the theoretical methods usually require lengthy derivations and complicated calculations of moments of the field. On the other hand, numerical simulations can provide an easier and more practical way to tackle beams with arbitrary amplitude, phase and polarization profiles. Numerical simulations can also be used to predict higher moments of the field, which can provide some basis for comparison with future analytic studies and experiments. Another drawback of analytical computations is that they only yield statistical quantities like the average irradiance and variance. However, in many applications like adaptive optics [17] and imaging [18], the amplitude and phase profile of the instantaneous optical beam is of interest rather than its statistical behavior. These systems require understanding of the evolution of the beam from one instant to another for an individual realization of the atmospheric turbulence. For such situations, numerically simulating the propagation of the beam is the preferred option. Simulations also provide a detailed beam profile as it propagates through the random medium, which is very useful for gauging the effect of the turbulence and the system performance. We want to further point out that most prior studies on beam propagation through atmosphere have dealt with scalar Gaussian beams. Numerical simulation methods described in this chapter can be beneficial for understanding and visualizing arbitrary structured beams (e.g. beams carrying OAM or polarization singularities) through realistic turbulence levels.

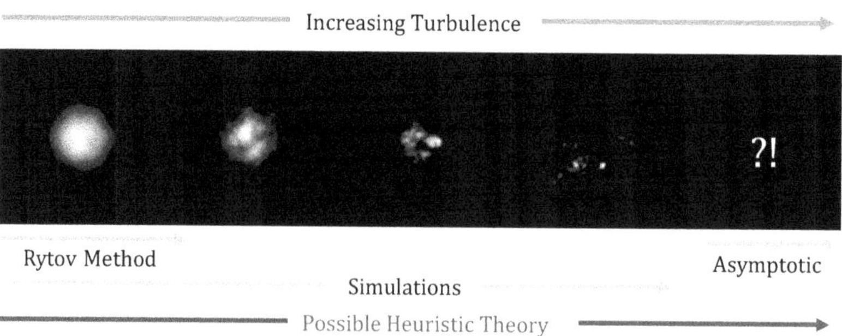


**Figure 8.1.** The range of validity of the Rytov method, numerical simulation studies, and asymptotic solutions with respect to the increasing turbulence strength is illustrated. In the absence of any heuristic theory, which can describe the experimentally observed results, simulation studies offer a practical solution. Figure layout adapted from [9]. Beam profiles are simulated with the numerical methods described later in this chapter.

The beam propagation studies through the atmosphere usually undertake one of two approaches—(a) the corpuscular [19] and (b) the wave approach [20]. In the corpuscular approach, the propagation of light is considered a random process of photon scattering by air molecules. This approach involves generating an ensemble of calculated photon trajectories, which can then be used to find the angular distribution, the polarization of the scattered beam, etc [19]. The second approach is the wave approach, which uses multiple phase screens to mimic the role of the atmosphere by perturbing the phase of the propagating wave [21–23]. Here, the beam propagation is considered a process of successive scattering of the light waves by these phase screens which have been generated by filtering Gaussian random noise with the atmospheric phase spectrum. This random phase screen method, also known as the split-step propagation method, is widely used in propagation studies [9, 24–29] of waves in various types of random media particularly when one is interested in keeping track of the associated field quantities rather than just the total irradiance.

In what follows we describe various concepts involved in numerical simulation of atmospheric laser beam propagation using the split-step propagation method. The detailed discussion should make it easier for readers to follow (and possibly modify) the computer simulation code for the split-step method given in appendix A. The extended atmosphere is represented by stacking a finite number of thin phase screens along the propagation direction, separated by carefully chosen distance $\delta z$ in free space depending on the turbulence strength. The essential details like the sampling criterion for the simulation grid, propagation path, number of phase screens to be used, etc, are described. The chapter also discusses some recent methods used for generating random phase screens and their limitations.

## 8.1 Split-step propagation method

The split-step propagation method is a popular approach for modeling electro-magnetic wave propagation through the atmosphere by numerically solving the parabolic wave equation [2]. This method provides a full-forward wave calculation with relatively few approximations and accommodates both vertical and lateral refractive index inhomogeneities in the atmosphere's refractive index [30–32]. This Fourier based split-step method belongs to the family of pseudo-spectral methods. It is widely used in computational mathematics to solve the time-dependent non-linear partial differential equations like the non-linear Schrödinger equation [33].

The parabolic wave equation method was first used to study the long-range propagation of radio waves in the troposphere in 1946 by Leontovich and Fock [34]. However, it was soon observed that this analytical method was not suitable for dealing with complicated atmospheric conditions. It was only after Hardin and Tappert introduced an algorithm based on the Fourier split-screen method for numerically solving the parabolic equation that this method gained popularity in both acoustics and atmospheric wave propagation studies [35–37]. The split-step method discussed by Hardin and Tappert can be applied to differential equations with the property that the terms with the highest derivatives are linear with constant coefficients, and the non-linear or variable coefficient terms have fewer or no derivatives [35, 38]. The essence of this method is to alternate between the following two steps [35]: (i) advance the solution using only the non-linear or variable coefficient term using an implicit finite difference approximation and (ii) advance the solution exactly using only the linear, constant-coefficient term through the FFT. The split-step formulation for the optical parabolic wave equation is described in the next section.

### 8.1.1 Split-step formulation from the parabolic equation

Consider the parabolic equation for wave propagation through atmosphere that may be obtained by neglecting the second derivative of the field in $z$:

$$\nabla_T^2 U + 2k^2 n_1 U + 2ik\frac{\partial U}{\partial z} = 0. \tag{8.1}$$

Here $\nabla_T^2 = \frac{\partial^2}{\partial x^2} + \frac{\partial^2}{\partial y^2}$ represents the transverse Laplacian operator. This equation can be rearranged as,

$$\frac{\partial U}{\partial z} = \left[\frac{i}{2k}\nabla_T^2 + ikn_1\right]U. \tag{8.2}$$

In order to apply the split-step formulation to this equation, it is necessary that the refractive index inhomogeneities of the medium satisfy the Markov assumption. In other words, the local index fluctuations $n_1(x, y, z)$ are delta correlated (or un-correlated between two $z$-slices) along the wave propagation direction. Consider the field $U(x, y, z)$ as it propagates between two planes, one located at $z_1 = z_o$ and the other at $z_2 = z_o + \delta z$ where $\delta z$ is chosen to be larger then the outer scale of the refractive index inhomogeneities but small enough to apply any perturbation method locally. It is then assumed that the field $U(x, y, z_2)$ can be calculated from the field $U(x, y, z_1)$ by adding small perturbations to this field locally [10]. This permits us to express the propagated field $U(x, y, z_2)$ in terms of $U(x, y, z_1)$ as:

$$U(x, y, z_o + \delta z) = \exp\left[\frac{i}{2k}\delta z\, \nabla_T^2 + ik\int_0^{\delta z} n_1(x, y, z)dz\right]U(x, y, z_o). \tag{8.3}$$

The first term in the exponent refers to the effect of the transverse field derivatives while the second term refers to the effect of the refractive index fluctuations. By making use of the Markov approximation, these two terms in the exponent can be separated with negligible error [2]. Equation (8.3) can be approximated analytically with second-order accuracy by the split operator:

$$U(x, y, z_o + \delta z) = \exp\left[ik\int_0^{\delta z} n_1(x, y, z)dz\right]\exp\left[\frac{i}{2k}\delta z\nabla_T^2\right]U(x, y, z_o). \tag{8.4}$$

This is the basic philosophy behind the split-step propagation method which allows one to numerically solve the parabolic wave equation. It can be seen from equation

(8.4) that the propagation of field through the atmosphere is reduced to two independent steps:

1. Propagation through free space to a distance $\delta z$ using the propagator:

$$h_f(x, y, z) = \exp\left[\frac{i}{2k}\delta z \nabla_T^2\right]. \tag{8.5}$$

2. Multiplying the resultant field by a phase function that represents the effect of the refractive index fluctuations of the medium over the same distance $\delta_z$:

$$t(x, y) = \exp\left[i\theta(x, y)\right] = \exp\left[ik\int_0^{\delta z} n_1(x, y, z)dz\right]. \tag{8.6}$$

This step is equivalent to multiplying the field with a random phase screen with transmission function $t(x, y, z) = \exp[i\theta(x, y)]$. Therefore, the split-step method is also known as the random phase screen method, where the atmosphere is described by a finite number of thin phase screens stacked along the propagation direction and separated by equal distance $\delta z$ in free space. These phase screens mimic the atmosphere by perturbing the phase of the propagating wave according to the atmospheric phase spectrum.

The first step of the split-step algorithm involves propagation of the field through free space using the propagator given in equation (8.5). For this step, the fluctuating part of refractive index $n_1(x, y) = 0$, so that

$$\frac{\partial U}{\partial z} = \frac{i}{2k}\nabla_T^2 U. \tag{8.7}$$

One can obtain the Fourier transform of this equation by using the following identities [2]:

$$\mathcal{F}\left[\frac{\partial g(x)}{\partial x}\right] = i2\pi f_x G(f_x), \tag{8.8}$$

$$\mathcal{F}\left[\frac{\partial^2 g(x)}{\partial x^2}\right] = -(2\pi f_x)^2 G(f_x), \tag{8.9}$$

where $G(f_x)$ is the Fourier transform of the function $g(x)$. This gives us the Fourier transform space,

$$\frac{\partial \tilde{U}}{\partial z} = i\pi\lambda(f_x^2 + f_y^2)\tilde{U}, \tag{8.10}$$

where $\tilde{U} = \mathcal{F}[U]$. This equation can be integrated over $z$ to obtain the free space propagation in the Fourier transform space as,

$$\tilde{U}(f_x, f_y, z_o + \delta z) = H_f(f_x, f_y)\tilde{U}(f_x, f_y, z_o) \tag{8.11}$$

where $H_f(f_x, f_y) = \exp\left[i\pi\lambda\delta z(f_x^2 + f_y^2)\right]$ is the Fresnel transfer function. Therefore, using the transfer function notation, the field $U(x, y, z_o + \delta z)$ is given by [2, 9]

$$U(x, y, z_o + \delta z) = \exp\left[ik\int_0^{\delta z} n_1(x, y, z)dz\right]$$
$$\times \mathcal{F}^{-1}\left[\exp(i\pi\lambda\delta z[f_x^2 + f_y^2])\tilde{U}(f_x, f_y, z_o)\right].$$

(8.12)

The Fresnel transfer function $H_f(f_x, f_y)$ is commonly referred to as a 'chirp' function because the absolute value of its phase increases as the square of the spatial frequency variable [39, 40]. Such chirp functions are not band-limited and thus requires careful choice of sampling conditions for aliasing-free simulation using FFT routines [23, 40, 41]. The phase screens used in atmospheric propagation simulations have their own set of sampling criterion, thus making it more challenging to do error-free Fresnel diffraction based calculations. Though many algorithms have been developed to carry out fast Fresnel diffraction calculations [23, 42], in this book, we will refrain from using them. Instead we will use the angular spectrum method for free space propagation as explained thoroughly in chapter 3. Such an approach helps us in two ways: (a) we use the full Helmholtz wave equation for the free space propagation and (b) the sampling requirements are somewhat simpler. Therefore, the field $U(x, y, z_o + \delta z)$ using the angular spectrum method is now given by:

$$U(x, y, z_o + \delta z) = \exp\left[ik\int_0^{\delta z} n_1(x, y, z)dz\right]$$
$$\times \mathcal{F}^{-1}\left[\exp(i\delta z\sqrt{k^2 - 4\pi^2(f_x^2 + f_y^2)})\tilde{U}(f_x, f_y, z_o)\right].$$

(8.13)

The Fourier based split-step algorithms provide an efficient way to numerically propagate the field through atmosphere. The implementation of the split-step method is described in detail in the next section.

## 8.1.2 Implementation of the split-step propagation method

The geometry for wave propagation using the split-step method is illustrated in figure 8.2. The random phase screens are separated by distance $\delta z$ and are placed at positions $(m - 1/2)\delta z$ with $m = 1, 2, 3, \ldots, M$ along the propagation direction where $M$ is the total number of screens used. The phase screens carry the integrated phase due to turbulence over the distance $\delta z$. This random phase is obtained by integrating the refractive index fluctuation $n_1$ over the distance $\delta z$ (see equation (8.6)). The field that is incident upon a screen is multiplied with the transparency $t(x, y)$ of that screen and then propagated to the next screen. This process is easily implemented using angular spectrum method. The two-dimensional Fourier transform of the field $U(x, y, z)$ is first multiplied with the free space transfer function $\alpha$ for propagation over distance $\delta z$:

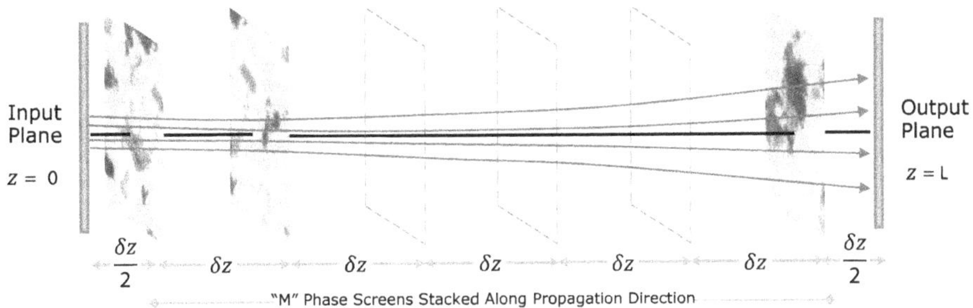


**Figure 8.2.** Schematic showing the placement of random phase screens in the split-step method for propagation of beams through atmospheric turbulence. Adapted with permission from [43] ©The Optical Society.

$$\alpha(f_x, f_y) = e^{i\delta z\sqrt{k^2 - 4\pi^2(f_x^2 + f_y^2)}}, \tag{8.14}$$

and the product is then inverse Fourier transformed to obtain the propagated field $U(x, y, z + \delta z)$. This propagated field now illuminates the next screen and the above process continues till the final distance $L$ is reached. The required steps of the propagation algorithm are shown in the flowchart of figure 8.3.

The accurate representation of the atmosphere and the free space diffraction of the beam depends on various important factors like selecting the simulation grid points and minimization the boundary effects in FFT based propagation routines, adequate sampling of the propagation path, and accuracy of phase reproduction on the random screens. The effect of these factors on the simulation accuracy can be minimized by fulfilling appropriate sampling requirements. The critical sampling criteria are described in detail in the following discussion.

### 8.1.3 Sampling requirements

Various sampling requirements need to be taken into account for accurate simulation of the wave propagation. These are described below:

1. **Simulation grid parameters**

    In order to get rid of aliasing error in the FFT based propagation method, it is necessary to choose the spatial grid sampling ($\Delta x = \Delta y$) and spatial frequency grid spacing ($\Delta f_x = \Delta f_y$) according to the Nyquist sampling requirements. Additionally, it may be required for an expanding beam to change the size of the computational window or sampling interval at different distances. Following the discussion in [40], the ideal sampling criterion for the free space transfer function given in equation (8.14) is

$$\Delta x = \frac{\lambda\sqrt{(L^2 + (D/2)^2)}}{D}, \tag{8.15}$$

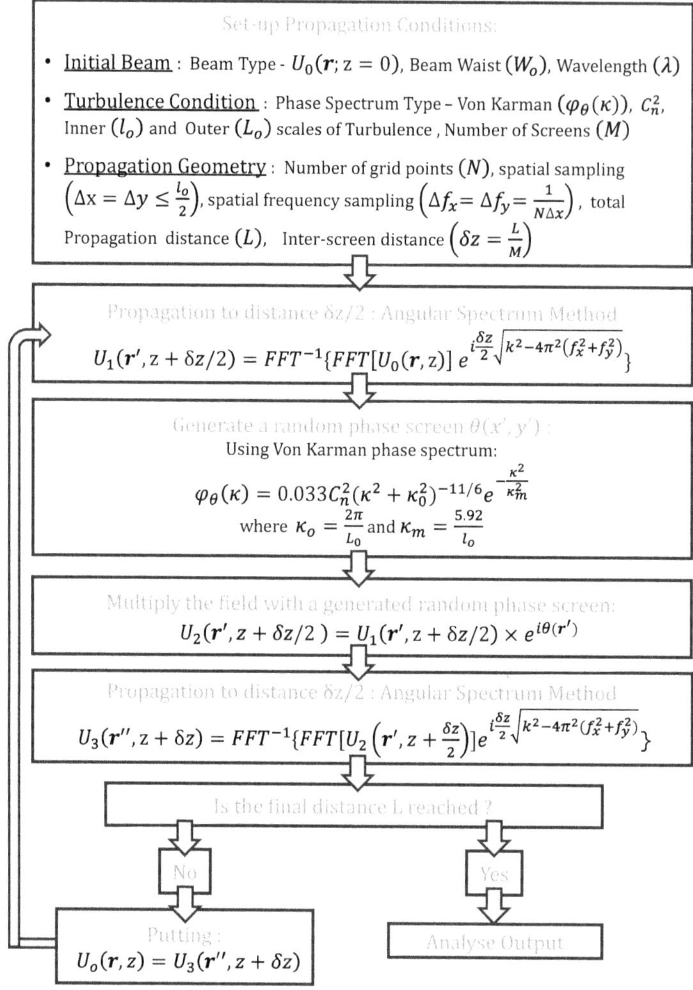


**Figure 8.3.** Flowchart denoting the steps for numerically propagating the beam through atmospheric turbulence using the random phase screen method.

where $L$ is the total propagation length and $D = N\Delta x$ is the total spatial size of the simulation grid. Here $N$ denotes the number of grid points. For paraxially propagating beams or when $L \gg D$, the sampling criterion simplifies to,

$$\Delta x = \frac{\lambda L}{D} = \sqrt{\frac{\lambda L}{N}}. \tag{8.16}$$

From the discussion of the free space transfer function $\alpha$ in chapter 3, we know that spatial frequencies higher than $1/\lambda$ cannot propagate in free space. By choosing the spatial grid sampling period $\Delta x > \lambda/2$, the maximum allowed spatial frequency on the grid becomes: $f_{\max} = \frac{1}{2\Delta x} < \frac{1}{\lambda}$. This ensures that one is only dealing with propagating spatial frequencies.

In case the computation needs to be performed with a larger sampling grid spacing $\Delta x'(=\Delta y')$, a low-pass filter defined as:

$$\mathrm{LP}(f_x, f_y) = 1, f_x^2 + f_y^2 < = f_0^2, \tag{8.17}$$

and zero elsewhere, needs to be used additionally for aliasing-free propagation computation [44]. Here the quantity $f_0$ representing the radius of the low-pass filter is given as:

$$f_0 = \frac{1}{\lambda\sqrt{1 + (2z/N\Delta x')^2}}. \tag{8.18}$$

2. **Accurate dynamic range**

One important limitation in the applicability of numerical simulations is the finite spatial dynamic range imposed by the maximum available numerical grid size. For a given simulation grid, the maximum and minimum allowed wave-numbers are $\kappa_{\max}^G = \frac{\pi}{\Delta x}$ and $\kappa_{\min}^G = \frac{\pi}{N\Delta x}$ respectively, with the available dynamic range,

$$\frac{\kappa_{\max}^G}{\kappa_{\min}^G} = N. \tag{8.19}$$

The spectral range defined by the numerical simulation parameters $[\kappa_{\max}^G, \kappa_{\min}^G]$ must provide sufficient bandwidth to include all the essential components of the spatial spectrum of the intensity fluctuations which are produced on propagation through turbulence. In this regard, the asymptotic solutions for spherical wave propagation in turbulence can be used to provide the required range of correlation scales of the intensity fluctuations. Though the asymptotic theory is valid only for strong path integrated turbulence, it is useful in defining the extreme limits to the simulation spectrum requirements.

Two important correlation scales can be defined for the intensity fluctuations, $d_o$ and $D_o$ as shown in figure 8.4. These two scales $d_o$ and $D_o$ are associated with the diffractive and refractive effects of the atmospheric turbulent eddies respectively. The atmosphere contains turbulent eddies of different sizes ranging from the outer scale $L_o$ to the inner scale $l_o$. These eddies depending on their spatial scales have different effects on the propagating wave. The larger scale eddies produce *refractive* effects which cause small-angle scattering and focusing resulting in beam wandering, whereas the smaller scale eddies produce *diffractive* effects which produce amplitude fluctuations and beam spreading. The small-scale $d_o \simeq r_{os}$ describes the feature size of the speckle pattern superimposed on the beam's irradiance profile. Here $r_{os}$ denotes the Fried parameter for spherical waves given by the expression,

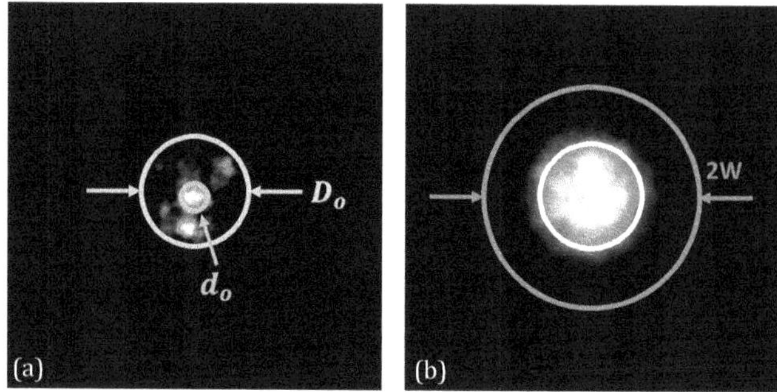



**Figure 8.4.** The short-time (instantaneous) and long-time averaged beam spot of a initially collimated Gaussian beam of 1.55 $\mu$m wavelength are shown in (a) and (b). The beam has been propagated over 2 km through moderate atmospheric turbulence given by $C_n^2 = 10^{-14}$ m$^{-2/3}$. Here, the diffractive short scale and refractive long scale are denoted by $d_o$ and $D_o$ respectively. The white circle shows the corresponding free space beam spot diameter.

$$r_{os} = \left[ 0.42k^2 \int_0^L C_n^2(z)\left(\frac{z}{L}\right)^{5/3} dz \right]^{-3/5}. \tag{8.20}$$

The larger scale $D_o$ originates in the strong scattering regime and is known as the effective beam spreading diameter. In isotropic turbulence, the scale $D_o$ has approximately the width of the scattering disk size that is, $D_o = 4L/kr_{os}$. The intensity is therefore modeled as a diffractive process of scale $d_o$ modulated by a larger refractive process of scale $D_o$. The intensity spectrum is thus band-limited by a minimum $\kappa_{min}^S = \pi/D_o$ and maximum spatial wavenumber $\kappa_{max}^S = \pi/d_o$. The required dynamic range for propagation simulation of spherical waves through turbulence is equal to

$$\frac{\kappa_{max}^S}{\kappa_{min}^S} = \frac{D_o}{d_o} = \frac{4L}{kr_{os}^2}. \tag{8.21}$$

For proper representation of all the required spatial frequencies in the simulation,

$$\frac{\kappa_{max}^G}{\kappa_{min}^G} \geqslant n\frac{\kappa_{max}^S}{\kappa_{min}^S}, \tag{8.22}$$

where $n$ is an integer $\geqslant 4$ for safely avoiding any aliasing errors. The necessary sampling condition from equations (8.19) and (8.20) can be stated as:

$$N \geqslant n\frac{4L}{kr_{os}^2}. \tag{8.23}$$

### 3. Beam broadening considerations

Consider a Gaussian beam with its field in $z = 0$ plane given by:

$$\psi(r; z = 0) = A \exp\left(-\frac{r^2}{W_o^2} - \frac{i\pi r^2}{\lambda F_o}\right), \tag{8.24}$$

where $A$ and $F_o (>0)$ denote the complex amplitude and the radius of curvature of the wavefront. Here, $W_o$ denotes the beam waist (or the $1/e^2$ beam intensity radius) at the $z = 0$. In free space, the beam waist at a distance $z > 0$ can be computed using Fresnel propagation and is given by the expression:

$$W(z) = W_o \sqrt{\left(1 - \frac{z}{F_o}\right)^2 + \left(\frac{z\lambda}{\pi W_o^2}\right)^2}. \tag{8.25}$$

However, in a turbulent medium, the propagating wave not only expands due to diffraction but also experiences a turbulence induced beam spreading. Turbulent eddies larger than the beam diameter deflect the beam spot and on propagation over larger distances, this deflection can become a serious issue as now the beam no longer falls onto the detector. The instantaneous beam width is given by the short-term beam radius defined as $W_s(z)$. The long-time averaged beam radius is given by:

$$W^2(z) = W_s^2(z) + 2\langle\beta^2(z)\rangle, \tag{8.26}$$

where $\beta(z)$ denotes the beam wander defined by the position of the beam centroid. Beam wander establishes the difference between the short- and long-term beam radius and is usually given as,

$$\beta(z) = [\beta_x(z), \beta_y(z)] = \frac{\iint \rho U(\rho, z)d\rho}{\iint U(\rho, z)d\rho}, \tag{8.27}$$

and

$$\langle\beta(z)^2\rangle = \langle\beta_x^2\rangle + \langle\beta_y^2\rangle. \tag{8.28}$$

Here, $U(\rho, z) = U(x, y, z)$ is the instantaneous beam intensity profile in the plane $z$. The short- and long-term averaged beam spots are shown in figure 8.4.

It is necessary to consider the relation between the beam size and the lateral dimension of the simulation grid ($N\Delta x$) so that the broadened beam is properly contained inside the simulation grid area for all propagation distances of interest. An approximate relation for the long-averaged beam radius in a turbulent medium was obtained by Yura [45] using the extended Huygens–Fresnel principle as:

$$W^2(z) = W_o^2\left[\left(1 - \frac{z}{F_o}\right)^2 + \left(\frac{\lambda z}{\pi W_o^2}\right)^2\right] + \left(\frac{D_o}{2}\right)^2. \tag{8.29}$$

The spatial extent of the simulation grid should be chosen such that

$$N\Delta x \geqslant 2mW(z), \tag{8.30}$$

where $m$ is an integer. The power contained in the tail of the Gaussian beam can be neglected outside a circular area of diameter $2W(z)$. Therefore, by choosing $m \geqslant 3$, it is ensured that the beam is contained within the simulation grid as it propagates, and no edge effects creep up. The other means of achieving this is by imposing absorption boundaries or using windowed illumination fields like super-Gaussian beams.

4. **Accurate phase representation and sampling of turbulence parameters**

   The phase screens used in the split-step propagation method also need to fulfill some numerical requirements for representing the atmospheric phase fluctuations accurately. The statistics of the beam intensity are susceptible to the small-scale phase fluctuations. So an important criterion for phase screens in this respect is that the inner scale of turbulence should be sampled appropriately. This is ensured by choosing $\Delta x \leqslant l_o/2$ according to the Nyquist sampling. Besides this, the change in the phase of the phase screen from one grid point to the other should be less than $\pi$ again to fulfill Nyquist sampling,

$$\theta(x_2, y_1) - \theta(x_1, y_1) < \pi, \tag{8.31}$$

or in terms of $\Delta x = x_2 - x_1$ and $\Delta y = y_2 - y_1$, we have

$$\Delta x \left|\frac{\partial \theta(x, y)}{\partial x}\right| < \pi, \tag{8.32}$$

and

$$\Delta y \left|\frac{\partial \theta(x, y)}{\partial y}\right| < \pi. \tag{8.33}$$

5. **Number of random phase screens required for given turbulence strength**

   The propagation path $L$ also needs to be sufficiently sampled in the $z$-direction. This is usually achieved by following the rule of thumb that less than 10% of the total intensity scintillation takes place over the inter-screen distance $\delta z$, that is, $\sigma_I^2(\delta z) < 0.1\,\sigma_I^2(L)$ where $\sigma_I^2(z)$ gives the intensity variance at any $z$ plane. The Rytov variance $(\sigma_R^2)$ is used as an estimate for intensity variance. It is further required that the intensity scintillation is weak over the distance $\delta z$. This can be ensured by placing a requirement that $\sigma_I^2(\delta z) < 0.1$. These two conditions can be used for determining the minimum number of screens required for any given value of $C_n^2$ and their placement along the propagation path. An estimate for obtaining the minimum number of phase screens has been provided in [46]:

$$M > (10\sigma_R^2)^{6/11}. \tag{8.34}$$

The phase screens should also be delta correlated along the $z$-direction as per the Markov approximation. For this, $\delta z$ must be chosen to be higher than the outer scale of turbulence $L_o$.

All the above conditions are essential for accurate modeling of the atmospheric wave propagation. In the next section, we will study the various methods for generating random phase screens in detail.

## 8.2 Phase screen generation

The simplest random phase screen model made use of only one phase screen and was first used in the analysis of wave diffraction in a thick slab of random inhomogeneous medium [47]. Such a treatment is useful when considering beam propagation through ionosphere or interplanetary plasma [48]. However, a single phase screen is not sufficient to describe the propagation of beams through extended turbulence, for example, the beam propagation over a horizontal propagation path over a few km in atmosphere. The error arising due to the representation of a continuum random medium by a single phase screen was studied by Booker *et al* [49]. It was observed that when the propagation length exceeds the outer scale ($z > L_o$), the single phase screen model significantly overestimates intensity fluctuations. In this case, a multiple phase screen model offers a more reliable representation of the turbulent medium [50].

The ability to generate accurate random phase screens lies at the core of the simulation for studying atmospheric turbulence effects. In order to simulate realistic turbulence phase fluctuations, the ensemble statistics of the random phase screens should match with those predicted by the theory. A large number of methods have been developed for phase screen generation. These methods can be broadly classified into two main types. The first type of phase screen generation method makes use of the Fourier transform to represent the phase screen as a two-dimensional rectangular grid of points by filtering complex Gaussian random numbers with the desired power spectral density of the phase fluctuations. The second type of phase screen generation method represents the phase screen as a sum of orthogonal basis functions [18, 51, 52]. In this section, we will concern ourselves with the FFT based phase screen generation method.

The atmospheric phase is a random function of the refractive index fluctuations. The atmospheric phase spectrum $\Phi_\theta(\kappa)$ corresponding to the turbulence induced phase fluctuations can be obtained from the power spectrum of refractive index fluctuations $\Phi_n(\kappa)$ as described in the following section.

### 8.2.1 Phase spectrum from the refractive index spectrum

Consider a thin slab of random medium of thickness $\delta z$ whose refractive index fluctuations are given by $n_1(r, z)$. From equation (8.6), the two-dimensional phase fluctuations $\theta(r)$ on passing through this slab are given by:

$$\theta(r) = k \int_0^{\delta z} n_1(r, z)dz. \tag{8.35}$$

The correlation function for phase can thus be written as:

$$\Gamma_\theta(r_1, z_1; r_2, z_2) = \langle \theta(r_1, z_1)\theta(r_2, z_2) \rangle$$
$$= k^2 \iint_0^{\delta z} \langle n_1(r_1, z_1)n_1(r_2, z_2) \rangle dz_1 dz_2, \tag{8.36}$$

where $\langle \ \rangle$ denotes ensemble average. The quantity $\langle n_1(r_1, z_1)n_1(r_2, z_2) \rangle$ is the correlation function $\Gamma_n(r_1, z_1; r_2, z_2)$ of the refractive index fluctuations. Under the assumption that $\delta z$ is much larger than the correlation length of the refractive index irregularities, one can make use of the Markov approximation to write $\Gamma_n(r_1, z_1; r_2, z_2)$ as

$$\Gamma_n(r_1, z_1; r_2, z_2) = \langle n_1(r_1, z_1)n_1(r_2, z_2) \rangle = \delta(z_1 - z_2)F_n(r_1 - r_2), \tag{8.37}$$

where $F_n(r_1 - r_2)$ is the two-dimensional power spectrum of the refractive index fluctuations in the transverse $x$-$y$ plane. The function $F_n(r_1 - r_2)$ may be obtained from $\Phi_n(\kappa)$ by integrating over the $z$ plane, that is,

$$F_n(r) = 2\pi \iint_{-\infty}^{+\infty} \Phi_n(\kappa_x, \kappa_y, \kappa_z = 0)\exp(i\kappa \cdot r)d\kappa_x d\kappa_y, \tag{8.38}$$

where $r = r_1 - r_2$. Substituting equation (8.37) into equation (8.36), one can write,

$$\Gamma_\theta(r_1, z_1; r_2, z_2) = k^2 \iint_0^{\delta z} \delta(z_1 - z_2)F_n(r)dz_1 dz_2 = k^2 \delta z\, F_n(r). \tag{8.39}$$

Next, we define a power spectral density $\Phi_\theta(\kappa)$ associated with the random phase fluctuations which forms a Fourier transform pair (Wiener–Khintchine theorem) with the phase correlation function $\Gamma_\theta(r)$, such that

$$\Phi_\theta(\kappa) = \int dr\, \Gamma_\theta(r)\exp(-i\kappa \cdot r). \tag{8.40}$$

Now, using equations (8.39) and (8.40), we see that the phase spectrum $\Phi_\theta(\kappa)$ takes the form,

$$\Phi_\theta(\kappa) = 2\pi\, k^2\, \delta z\, \Phi_n(\kappa_x, \kappa_y, \kappa_z = 0). \tag{8.41}$$

This expression allows one to obtain the phase spectrum corresponding to a given refractive index spectrum $\Phi_n(\kappa)$.

The refractive index fluctuations can be modeled as a stochastic process. The phase $\theta(r)$ is essentially a sum over many $n_1$ values (see equation (8.35)) as $\delta z$ is taken to be larger than the correlation length of the turbulence. This enables us to model the phase $\theta(r)$ as a Gaussian random process as per the central limit theorem. As explained in chapter 2, the traditional method for generating realizations of a random field with a well-defined power spectrum involves filtering of Gaussian white noise with the square root of the desired spectrum, followed by an inverse Fourier transform operation. This method is known as the FFT method and is commonly used to generate random realizations of atmospheric turbulence in simulation studies. In the next section, we will describe the FFT method for phase screen generation in detail. A few other known methods used for random screen generation will also be briefly discussed in the end.

## 8.2.2 FFT method for phase screen generation

The topic of generating realizations of a random process with a given power spectrum was already discussed briefly in section 2.3. A continuous two-dimensional random phase screen $\theta(r)$ can be represented as a Fourier integral

$$\theta(r) = \iint_{-\infty}^{+\infty} g(\kappa_x, \kappa_y)\sqrt{\Phi_\theta(\kappa_x, \kappa_y)}\exp(i\kappa \cdot r)d\kappa_x d\kappa_y, \tag{8.42}$$

where $\kappa = [\kappa_x, \kappa_y, \kappa_z = 0]$ and $g(\kappa_x, \kappa_y)$ is a zero mean, unit-variance complex Gaussian white noise process. In order to simulate the phase screen on a numerical grid, equation (8.42) needs to be converted to its approximate discrete form. The discrete phase screen is written as a sum of Fourier harmonics with random complex coefficients,

$$\theta_{\mathrm{FFT}}(x, y) = \sum_{n,m\,=-N/2}^{N/2-1} g(\kappa_x, \kappa_y)\sqrt{\Phi_\theta(\kappa_x, \kappa_y)}$$
$$\times \exp\left[i2\pi(f_x x + f_y y)\right]\Delta\kappa_x\,\Delta\kappa_y. \tag{8.43}$$

This equation can be easily implemented using the numerically efficient Fast Fourier Transform (FFT) algorithm. In the above equation, $\kappa_x = 2\pi f_x$ and $\kappa_y = 2\pi f_y$. The sample points on the numerical grid in the spatial and frequency domain can be defined as $[x, y] = [j\Delta x, l\Delta y]$ and $[f_x, f_y] = [n\Delta f_x, m\Delta f_y]$ respectively. The sample intervals are chosen as $\Delta x = \Delta y$ and $\Delta f_x = \Delta f_y = 1/(N\Delta x)$ where $N$ is the total number of grid points. The function $g(\kappa_x, \kappa_y)$ denotes a discrete complex Gaussian noise process given by:

$$g(\kappa_x, \kappa_y) = \frac{g(n, m)}{\sqrt{\Delta\kappa_x\,\Delta\kappa_y}} = \frac{a(n, m) + ib(n, m)}{\sqrt{\Delta\kappa_x\,\Delta\kappa_y}}, \tag{8.44}$$

where the discrete random numbers $a(n, m)$ and $b(n, m)$ are drawn from a Gaussian random distribution with standard deviation equal to unity. Therefore, the expression for the simulated phase screen is given by:

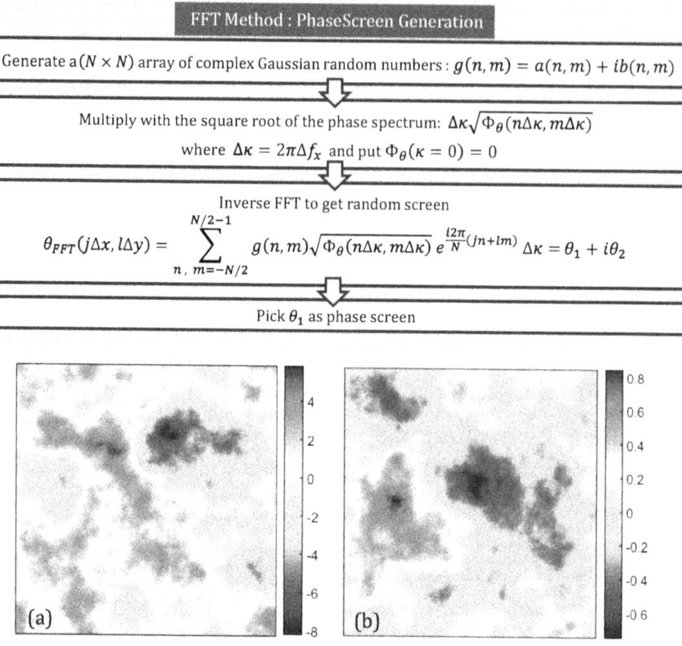



**Figure 8.5.** Algorithm for the FFT method for random phase screen generation. Here (a) and (b) show examples of generated phase screens when the structure function for the refractive index fluctuations $C_n^2 = 10^{-13}\,\mathrm{m}^{-2/3}$ and $C_n^2 = 10^{-15}\,\mathrm{m}^{-2/3}$ respectively. The above phase screens are generated using a 512 × 512 simulation grid with von Kárman phase spectrum having inner and outer scale equal to $l_o = 1$ cm, $L_o = 10$ m respectively and the propagation step length is taken as $\delta z = 25$ m.

$$\theta_{\mathrm{FFT}}(j\Delta x,\, l\Delta y) = \sum_{n,m=-N/2}^{N/2-1} g(n,\, m)\sqrt{\Phi_\theta(n\Delta\kappa,\, m\Delta\kappa)}$$
$$\times \exp\left[i\frac{2\pi}{N}(jn + lm)\right]\Delta\kappa. \tag{8.45}$$

The FFT operation above is expected to give a complex-valued function in general whose real part is typically used as the required phase function $\theta_{\mathrm{FFT}}$. The value of the phase spectrum $\Phi_\theta$ at the origin ($n = m = 0$) is usually set equal to zero. This has no effect on the spatial statistics of the wave-field as the zero-frequency component of the phase screen only determines the average phase delay offered by the screen. This term has no contribution to the tip, tilt or behavior of the propagating field. Besides, it has been shown that a large enough value of $\Phi_\theta(0, 0)$ may lead to a quantization error [53]. Figure 8.5 details the steps for the random screen generation using the FFT method. The generated sample phase screens for two different $C_n^2$ values: $C_n^2 = 10^{-13}\,\mathrm{m}^{-2/3}$ and $C_n^2 = 10^{-15}\,\mathrm{m}^{-2/3}$ respectively are also shown in figures 8.5(a) and (b). Note that the range of phase variation in the phase screens for a given $\delta z$ is dependent on the value of the structure constant $C_n^2$ of the refractive index fluctuations.

*Accuracy of the FFT based random screen simulator*
The accuracy of the phase screens generated by any method is typically evaluated by their ability to reproduce the desired phase structure function for the given turbulence model. The structure function for the phase fluctuations is defined as,

$$D_\theta(r) = \langle [\theta(r_1) - \theta(r_2)]^2 \rangle, \tag{8.46}$$

which represents the average squared difference in the phase of the screen for a pair of points which are separated by $r = r_1 - r_2$ from each other. The structure function is related to the two-dimensional phase correlation function $\Gamma_\theta(r)$ of the phase screen as,

$$D_\theta(r) = 2[\Gamma_\theta(0) - \Gamma_\theta(r)], \tag{8.47}$$

where

$$\Gamma_\theta(r) = \langle \theta(r_1)\theta(r_2) \rangle. \tag{8.48}$$

The two-dimensional phase correlation function is equal to the Fourier transform of the two-dimensional phase spectrum,

$$\Gamma_\theta(r) = \iint_{-\infty}^{+\infty} \Phi_\theta(\kappa_x, \kappa_y, \kappa_z = 0)\exp(i\kappa \cdot r)d\kappa_x d\kappa_y. \tag{8.49}$$

Therefore, a discrete form of two-dimensional correlation function for the FFT based phase screen can be obtained by taking an inverse FFT of the discrete two-dimensional phase spectrum[1] as,

$$\Gamma_{\theta_{FFT}}(j\Delta x, l\Delta y) = \sum_{n,m=-N/2}^{N/2-1} \Phi_\theta(n\Delta\kappa, m\Delta\kappa)(\Delta\kappa)^2$$
$$\times \exp\left[i\frac{2\pi}{N}(jn + lm)\right]. \tag{8.50}$$

The above expression for the two-dimensional correlation function can be substituted into equation (8.47) to calculate the expected structure function. It is observed that for a statistically large enough ensemble of random phase screens (>1000), the computed average structure function closely matches the expected structure function. The expected phase structure function for the random screens is compared with the theoretical structure function for the same atmospheric model and turbulence parameters. In general, the theoretical structure function for phase screens is given by

---

[1] The pole at the origin of the phase spectrum is usually handled by putting the numerical spectrum value equal to zero.

$$D_\theta(r) = 2 \iint_{-\infty}^{+\infty} \Phi_\theta(\kappa)[1 - \exp(i\kappa \cdot r)]d\kappa_x \, d\kappa_y. \tag{8.51}$$

For the case of modified von Karman turbulence spectrum, the theoretical phase structure function $D_\theta$ has the form:

$$D_\theta^{mvK}(r) = 3.08 \, r_{of}^{-5/3}$$

$$\times \left\{ \Gamma\left(-\frac{5}{6}\right)\kappa_m^{-5/3}\left[1 - {}_1F_1\left(-\frac{5}{6}; 1; -\frac{\kappa_m^2 r^2}{4}\right)\right] - \frac{9}{5}\kappa_0^{1/3}r^2 \right\} \tag{8.52}$$

where ${}_1F_1(a; c; z)$ is a confluent hypergeometric function of the first kind and $r_{of}$ represents the Fried parameter for plane wave,

$$r_{of} = \left[0.423k^2 \int_0^L C_n^2(z)dz\right]^{-3/5}. \tag{8.53}$$

This structure function expression can be written in a simpler form within <2% error margin by using an algebraic approximation for the hypergeometric function, which yields

$$D_\theta^{mvK}(r) \approx 7.75 \, r_{of}^{-5/3}l_o^{-1/3}r^2\left[\frac{1}{\left(1 + \dfrac{2.03r^2}{l_o^2}\right)^{1/6}} - 0.72(\kappa_o l_o)^{1/3}\right]. \tag{8.54}$$

Figure 8.6 compares the expected structure function for the FFT based phase screens and the theoretical structure function. Here, von Kárman phase spectrum is used with two different values of the outer scale $L_o = 10$ m and $L_o = 100$ m respectively in plots in figures 8.6(a) and (b). The other simulation parameters are: $N = 512$, $D = 2$ m, $l_o = 1$ cm, $C_n^2 = 10^{-14}$ m$^{-2/3}$ and $\delta z = 100$ m. The difference between the

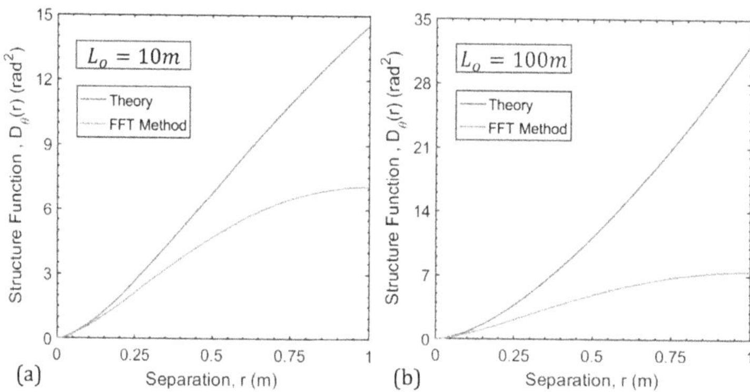


**Figure 8.6.** The structure function for the FFT based phase screens is compared with the theoretical structure function. The simulation parameters are detailed in the text.

theoretical structure function and the simulated structure function is readily apparent in the two plots. This big discrepancy in the behavior of the simulated structure function is due to the inadequate sampling of the low-frequency content of the two-dimensional phase spectrum by the FFT based simulation method.

The FFT based simulation method is thus seen to suffer from the basic limitation that the available bandwidth of spectral components generated in the phase screen is decided by the numerical grid parameters. From the perspective of modeling atmospheric turbulence, this means that the width of the phase screen ($D$) decides the simulated outer turbulence scale while the sampling size ($\Delta x$) decides the simulated inner turbulence scale of the phase screens. It is quite possible that the parameters $D$ and $\Delta x$ associated with the phase screen would not accurately represent the actual inner and outer scales of the atmospheric turbulence. The maximum and minimum spatial frequencies of the simulation are given by $f_{\min}^{\text{sim}} = \Delta f = 1/D$ and $f_{\max}^{\text{sim}} = N\Delta f/2 = N/2D$. As the frequencies lower than $f_{\max}^{\text{sim}}$ are neglected, this means that the phase screen does not incorporate the effects due to the refractive index fluctuations having sizes greater than the screen size.

The inner scale is not represented properly on the numerical grid (or the phase spectrum is under-sampled at high frequencies) when the sampling size $\Delta x$ is greater than the inner scale value $l_o$. Such cases are generally not severe due to the spectral roll-off at higher $\kappa$ values (note the exponential decay term in the von Kárman spectrum model). However, under-sampling of the outer scale puts a serious limit on the applicability and accuracy of the FFT based phase screens. The outer scale is the main contributor to the turbulence's low-frequency characteristics, for example, tilt. These lower frequencies are responsible for causing the wandering of the beam centroid, which may become significant for long propagation distances. The minimum and maximum spatial frequencies corresponding to the parameters used for simulating figure 8.6 are $f_{\min}^{\text{sim}} = 0.5$ m$^{-1}$ and $f_{\max}^{\text{sim}} = 128$ m$^{-1}$ which respectively corresponds to scale sizes of 2 m and 0.0078 m. Thus, the outer scale parameter $L_o = 10$ m in figure 8.6(a) and $L_o = 100$ m in figure 8.6(b) is not appropriately represented by the simulated screen. In order to properly sample the outer scale $L_o = 10$ m, the first sample away from origin in two-dimensional Fourier space should occur at $f_{\min} = 1/L_o = 0.1$ m$^{-1}$ whereas in the simulated screen, the first sample is located at $f_{\min}^{\text{sim}} = 0.5$ m$^{-1}$. Therefore, a major portion of the spectral energy near the origin is not sampled appropriately and the resultant phase screens therefore do not have accurate large-scale turbulence characteristics.

A straightforward way to incorporate lower spatial frequencies in the phase screen is by increasing the size of the grid by increasing $N$ and keeping $\Delta x$ the same. However, this will result in very large grid sizes that are not practically feasible. For example, a grid size, which is five times the outer scale value, has been suggested for adequately sampling the outer scale [54]. In that case, for $L_o = 10$ m, one would require $D = 50$ m or an impractically high $N = 12\,800$. Another scheme that has been explored in this regard is to cut out a small portion from a very large phase screen and use it in propagation simulations [26]. It is expected that the lower frequency characteristics will be accurate over a small portion of a large screen.

A Zernike polynomial based representation of the Kolmogorov's turbulence spectra was provided in [55]. One can add low order Zernike modes using Noll's method to compensate for the low frequencies. Recently, low-frequency compensation using correlation matrix based methods have also been proposed [56]. While the methodology described above leads to a square shaped phase screen, there are applications like laser beam tracking, where beam propagation through turbulence over a range of continuously varying directions is required. A phase screen that extends over large distances in one dimension then becomes desirable for simulation purposes. One straightforward approach for generating screens of this nature involves using rectangular windows in the FFT screen method where the spatial frequency sampling in two orthogonal directions needs to be adjusted appropriately. A computationally efficient 'infinite' random screen approach [57] has also been developed for cases when it is required to simulate a very long rectangular screen with slowly varying turbulence statistics. We will not describe this methodology in further details in this book.

In the next section, we discuss a more commonly used method for accounting for these lower frequencies. This method involves generation and addition of a sub-harmonic phase screen to the existing FFT based phase screen. Multiple methods have been proposed to incorporate the sub-harmonic frequencies in a phase screen [26, 54].

### 8.2.3 Sub-harmonic correction to the phase screen

An efficient way of adding low-frequency information without employing a very large-sized phase screen is through the sub-harmonic method [26, 54, 58–60]. The basic idea is to generate a separate low-frequency (sub-harmonic) phase screen using the FFT method that finely samples only the low-frequency portion of the phase spectrum. This sub-harmonic screen is then added to the phase screen (simulated as explained in section 8.2.2) for accurately modeling the large-scale atmospheric turbulence effects. A sub-harmonic is basically a sinusoidal function with a period larger than the screen size which has a spatial frequency $f^{\text{sub}}$ much lower than the lowest sampled frequency $f^{\text{sim}}_{\text{min}}$ of the phase screen. The sub-harmonic screen is generated by sampling the phase spectrum in the vicinity of $\kappa = 0$, at a much smaller sampling interval $f^{\text{sub}}$, than the rest of the spectrum. In this section, we will discuss the generation of this sub-harmonic phase screen as presented by Lane, Glindemann, and Dainty [58], which was later modified by Frehlich [59] to improve convergence.

The central pixel corresponding to spatial frequencies $f_x = 0$ and $f_y = 0$ of the high frequency phase screen is divided into nine equally sized sub-pixels as shown in figure 8.7(a). Note that each sub-pixel is 1/9 of the area of the original pixel size ($\Delta f_x$) and the sub-harmonic sample width is $\Delta f_x^{\text{sub}} = \Delta f_x / 3$. This describes the first sub-harmonic level. Sample points are now placed in the eight outer sub-pixels which are then used to generate the low-frequency screen using the FFT method. The process is now repeated again on the central small sub-pixel at the origin, creating multiple sub-harmonic levels as needed. The sample sizes of the sub-pixels at the $p$th

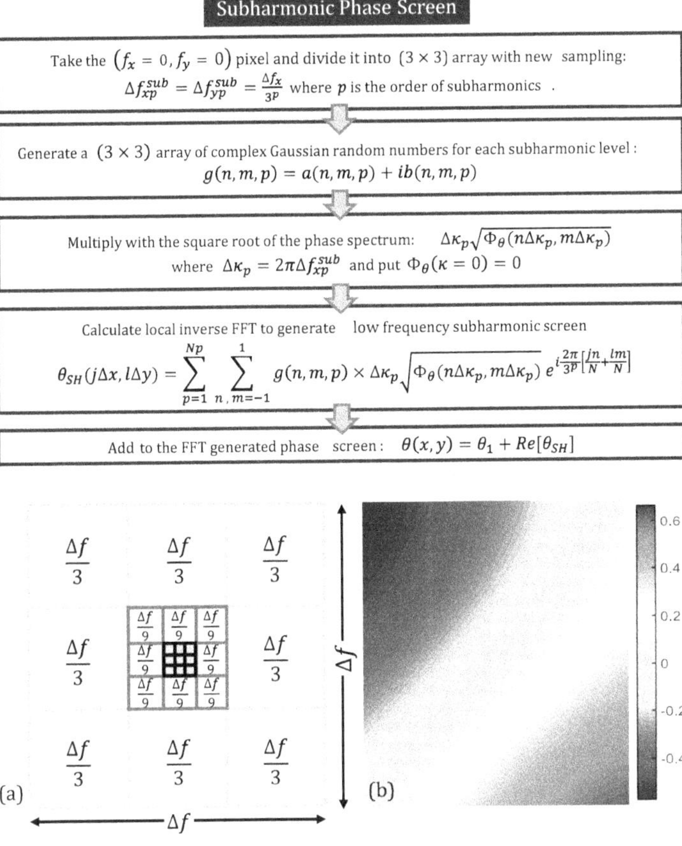





**Figure 8.7.** The flowchart lists the steps for generation of a sub-harmonic phase screen. The sub-harmonic grid used in calculation and a sample sub-harmonic phase screen ($\theta_{SH}$) for $C_n^2 = 10^{-14}$ m$^{-2/3}$ is shown in (a) and (b) respectively.

sub-harmonic level ($p \geqslant 1$) are given by $\Delta f_{xp}^{sub} = \Delta f_x/3^p$. The phase contributions from all the sub-harmonic levels are added together to generate the total sub-harmonic screen. Assuming $N_p$ sub-harmonic levels, the sub-harmonic phase screen is given by:

$$\theta_{SH}(j\Delta x, l\Delta y) = \sum_{p=1}^{Np} \sum_{n,m=-1}^{1} g(n, m, p)\Delta\kappa_p\sqrt{\Phi_\theta(n\Delta\kappa_p, m\Delta\kappa_p)} \\ \times \exp\left[i\frac{2\pi}{N3^p}(jn + lm)\right], \tag{8.55}$$

where $\Delta\kappa_p = 2\pi\Delta f_{xp}^{sub}$ and $g(n, m, p)$ denotes the complex Gaussian random number for the $(n, m)$ sub-pixel of $p$th sub-harmonic level. In the above summation, for each

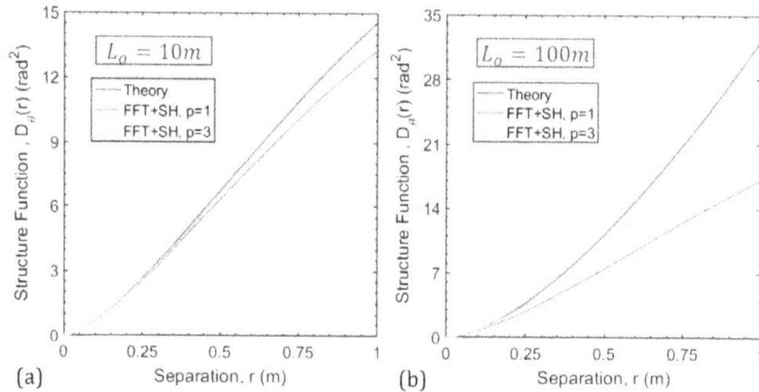

**Figure 8.8.** The expected structure function for the FFT method phase screens with two levels of sub-harmonic correction are compared with the theoretical structure function.

sub-harmonic level, the central sub-pixel ($m = n = 0$) representing the phase spectrum value at origin $\Phi_\theta(0, 0)$ is set equal to zero. The steps for generation of the sub-harmonic phase screen and a sample sub-harmonic screen are illustrated in figure 8.7. The final phase screen with accurate turbulence scales is then obtained by summing the sub-harmonic screen and the FFT based phase screen:

$$\theta_{\text{Total}}(j\Delta x, l\Delta y) = \theta_{\text{FFT}}(j\Delta x, l\Delta y) + \theta_{\text{SH}}(j\Delta x, l\Delta y). \tag{8.56}$$

Let us look at the accuracy of this phase screen by comparing its expected structure function with the theoretical value. The phases $\theta_{\text{FFT}}$ and $\theta_{\text{SH}}$ are mutually independent variables with zero mean, thus from equation (8.48), one can write the two-dimensional correlation function for $\theta_{\text{Total}}$ as [56],

$$\Gamma_{\theta_{\text{Total}}} = \Gamma_{\theta_{\text{FFT}}} + \Gamma_{\theta_{\text{SH}}}. \tag{8.57}$$

The discrete two-dimensional correlation function for the sub-harmonic phase screen is given by $\Gamma_{\theta_{\text{SH}}}$

$$\Gamma_{\theta_{\text{SH}}}(j\Delta x, l\Delta y) = \sum_{p=1}^{Np} \sum_{n,m=-1}^{1} \Phi_\theta(n\Delta\kappa_p, m\Delta\kappa_p)(\Delta\kappa_p)^2$$
$$\times \exp\left[ i\frac{2\pi}{N3^p}(jn + lm) \right]. \tag{8.58}$$

The expected structure function for phase $\theta_{\text{Total}}$ can be calculated from equations (8.47), (8.57) and (8.58) using the method outlined in section (8.2.2). Figure 8.8 compares the expected structure function to the theoretical structure function $D_\theta^{mvK}$ for the cases of when sub-harmonic screens up to levels $N_p = 1$ and $N_p = 3$ were used. This figure corresponds to the same simulation parameters used for testing the FFT method based phase screens as in figure 8.6. It can be observed that the expected structure function agrees closely with the theoretical structure function when sufficient number of sub-harmonic levels have been included, thus pointing towards a more accurate representation of the associated turbulence scales. The case $L_o = 10$ m, $p = 3$ gives a very close match with $D_\theta^{mvK}$, however, for $L_o = 100$ m, one needs to include more sub-harmonic levels $p > 3$. However, it has been observed that accuracy of the structure function generally saturates after a certain number of sub-harmonic levels. Adding sub-harmonic levels after $p = 15$ is usually not numerically feasible due to the overflow for double precision floating point calculations. Thus, for big (or infinite) outer scale values, the expected structure function is likely to deviate from its theoretical value at large separations. Banakh *et al* [61] studied the effectiveness of the sub-harmonic method for simulating large-scale turbulence inhomgeneities. They observed that with an increase in the number of the sub-harmonics, there was an increase in both the effective radius and relative variance of intensity fluctuations of the beam as calculated from simulation data. And these values had a better agreement with the theoretical and experimental data. Thus, inclusion of sub-harmonics resulted in more accurate estimate of the beam parameters. They further observed that increase in the number of sub-harmonic levels above $N_p = 8$ was ineffective as now each higher sub-harmonic level contributed weakly compared to the previous level. Another study by Sedmak [62], provided a rough estimate for the optimum number of sub-harmonic levels that should be used. It was observed that the number of sub-harmonic levels required to obtain the optimum structure function is proportional to the ratio of the turbulence outer scale length to the screen size. Carbillet and Riccardi [63] further defined two criteria based on the integrated power over the whole range of spatial frequencies and the structure function ratio which could be used to decide the minimum number of required sub-harmonics for atmospheric turbulence phase screen simulation. Another detailed investigation for optimal sub-harmonics selection appropriate for the Kolmogorov model, the von Kármán model and the modified von Kármán model was carried out by Liu *et al* [64]. The sub-harmonic method remains a widely popular method to overcome the inadequacies in the modeling of the low frequencies in the FFT based method by increasing the sampling density of the spatial frequencies in the vicinity of the origin.

### 8.2.4 Some drawbacks of the FFT method

The FFT based phase screen methods remain very popular owning to the computational efficiency of the FFT algorithm. However, there are certain disadvantages associated with this method:

1. **Under-sampling of the low spatial frequencies**:

    In the spatial domain, the phase is sampled on a fixed discrete rectangular grid. Therefore, one cannot exactly reproduce the desired phase structure function on a discrete grid as it ideally requires continuous values of the spatial frequencies $\kappa$. Besides, the central pixel corresponding to $n = m = 0$ in equation (8.45) is always put equal to zero, which leads to the under-sampling at low spatial frequencies. The contribution of these low spatial frequencies can be accounted for to a large extent by adding a sufficient number of sub-harmonics. However, in that case the FFT algorithm cannot be used for the calculation of the sub-harmonics, leading to an additional computational penalty. Also, the number of spectral components needed

$(N^2 + 8N_p)$ now exceeds the number of points $(N^2)$ in the spatial grid and could be very large for realistic phase screens.

2. **Periodic boundary conditions**: The phase screens are periodic with periods equal to the grid size $D$ due to the periodicity of the FFT algorithm. The periodic phase screens do not have a significant average slope [58] and so do not simulate the beam wandering correctly. This further causes difficulties in simulation of moving turbulence [65].

3. **Anisotropic phase screens**: For any rectangular (or square) phase screen, the loss of the low-frequency components in the diagonal direction is less than that in the $x$ or $y$ direction [56]. Thus the statistical properties of the FFT based phase screen are anisotropic in nature.

## 8.3 Other methods for generating random phase screens

Besides the FFT method, there are various other methods available in the literature which can be used to generate phase screens with good computational efficiency and accuracy [51, 66–75]. Goldring and Carlson[76] have reviewed five common approaches which use computational linear algebra and FFT for phase screen generation. Next, we will discuss two new recent approaches in this regard.

### 8.3.1 Randomized spectral sampling method

Recently, Paulson, Wu and Davis presented a new modified FFT method [77] for phase screen generation based on the frequency shift property of the Fourier transform [39]. This method easily incorporates the lower spatial frequency components in the phase screen without the additional computation time penalties associated with methods using sub-harmonic grids. They successfully demonstrated that for simulations of atmospheric turbulence with finite outer scales, the statistical phase structure function of the generated phase screens is in good agreement with theory. Additionally this method can be combined with the sub-harmonic method of Lane *et al* [58] to give very accurate results across a range of spectral models of practical and theoretical interest.

Consider a phase screen simulated using the FFT based method that is defined on a square grid as in equation (8.45). We rewrite this equation as,

$$\theta_{\text{FFT}}(j\Delta x,\, l\Delta y) = \sum_{n,m=-N/2}^{N/2-1} \tilde{c}(n\Delta\kappa,\, m\Delta\kappa)\exp\left[i\frac{2\pi}{N}(jn + lm)\right], \tag{8.59}$$

where

$$\tilde{c}(n\Delta\kappa,\, m\Delta\kappa) = g(n, m)\sqrt{\Phi_\theta(n\Delta\kappa,\, m\Delta\kappa)}\ \Delta\kappa. \tag{8.60}$$

Paulson *et al* [77] proposed a more meaningful use of the point closest to the $\kappa$-space origin by defining a new type of complex phase screen, $\theta_{\text{PWD}}$ such that:

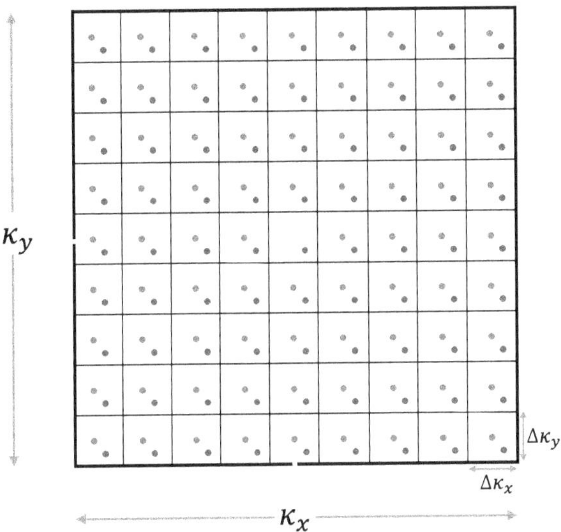


**Figure 8.9.** The $\kappa$-space grid sampling for the (a) traditional (blue dots) and (b) the randomized (magenta dots) spectral sampling methods is shown for one realization of the random phase screen. Note that in traditional FFT phase screens, the central pixel representing $\kappa(0, 0)$ is put to zero, whereas in the randomized spectral sampling method, a sampling point is present in the central pixel.

$$\theta_{\mathrm{PWD}}(j\Delta x,\, l\Delta y) = \sum_{n,m=-N/2}^{N/2-1} \tilde{c}(n\Delta\kappa + \delta\kappa_x,\, m\Delta\kappa + \delta\kappa_y)$$
$$\times \exp\left[i(j\Delta x(n\Delta\kappa + \delta\kappa_x) + l\Delta y(m\Delta\kappa + \delta\kappa_y))\right], \tag{8.61}$$

where $\delta\kappa_x$, $\delta\kappa_y$ are random variables described by a uniform distribution bounded by $\pm\Delta\kappa_x/2$ and $\pm\Delta\kappa_y/2$ respectively[2]. This offsets the lowest wave-number grid point away from the origin in addition to translating the rest of the sampling grid in the frequency domain. Figure 8.9 shows the difference between the $\kappa$-space grid partitioning and sampling for the traditional FFT spectral sampling approach versus the randomized spectral sampling approach. By defining a new matrix $C(j\Delta x,\, l\Delta y)$ with its elements related to $\tilde{c}(n\Delta\kappa + \delta\kappa_x,\, m\Delta\kappa + \delta\kappa_y)$ by Fourier transform, one can rewrite the expression in equation (8.61) as:

$$C(j\Delta x,\, l\Delta y) = N^2 . \, \mathcal{F}^{-1}\left[\tilde{c}(n\Delta\kappa + \delta\kappa_x,\, m\Delta\kappa + \delta\kappa_y)\right], \tag{8.62}$$

$$\theta_{\mathrm{PWD}}(j\Delta x,\, l\Delta y) = \exp[i(j\Delta x\delta\kappa_x + l\Delta y\delta\kappa_y)] \cdot C(j\Delta x,\, l\Delta y). \tag{8.63}$$

The complex phase screen $\theta_{\mathrm{PWD}}$ and the real-valued phase screens obtained from its real and imaginary parts, no longer exhibit periodicity and contain domain wide low spatial frequency distortions. Figure 8.10 shows the comparison between the phase

---

[2] Note that we have $\Delta\kappa_x = 2\pi/N\Delta x$ and $\Delta\kappa_x = \Delta\kappa_y$. So, we replace $2\pi/N = \Delta\kappa\Delta x = \Delta\kappa\Delta y$ in equation (8.61).

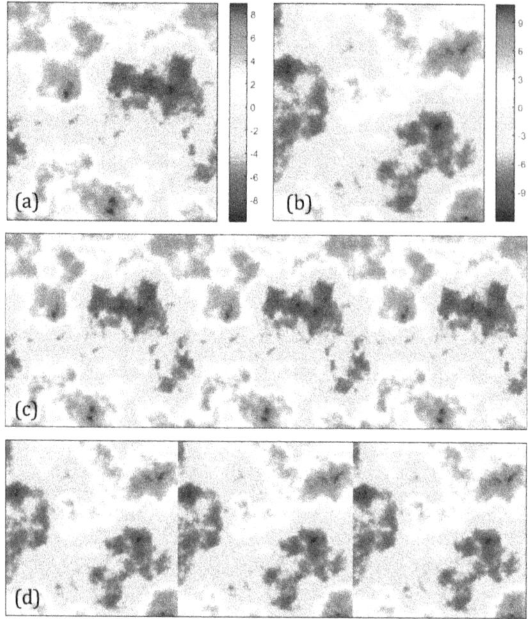

**Figure 8.10.** Sample phase screens are shown for the (a) traditional FFT method and (b) the modified FFT method based on randomized spectral sampling. The above phase screens are generated using a $512 \times 512$ simulation grid with von Kárman phase spectrum for $C_n^2 = 10^{-13}\,\mathrm{m}^{-2/3}$, inner and outer scale equal to $l_o = 1$ cm, $L_o = 3$ m respectively and the propagation step length $\delta z = 50$ m. To illustrate the periodic and aperiodic nature of these two types of phase screens shown in (a) and (b), we have stacked three identical copies of these phase screens adjacent to each other in (c) and (d) respectively. The aperiodic nature of the phase screens generated by using the new modified FFT method is clearly visible in (d).

screens obtained from traditional FFT method and the modified FFT method. The usual convention of setting $\tilde{c}(0, 0)$ to zero is due to the lack of a traditional spectral sampling grid point at the origin. The present method, however, provides a more meaningful use of the point closest to the $\kappa$-space origin. On further investigation, it was observed that for structure function power laws greater than the 2/3 law for Kolmogorov spectrum, the randomized spectral sampling algorithm alone was actually not sufficient to ensure accurate statistics of observed simulated structure function. For this, a new hybrid algorithm utilizing both FFT based frequency sampling randomization and sub-harmonic frequency sampling randomization was proposed [77, 78].

### 8.3.2 Sparse spectrum method

The sparse spectrum (SS) model was first proposed by M Charnotskii for modeling sea surface elevation [79]. This technique was later introduced as an efficient way of simulating the phase front perturbations caused by atmospheric turbulence [78, 80, 81]. This model preserves the wide range of scales typically associated with atmospheric turbulence perturbations with relatively less computational effort.

The technique is based on the assumption that individual phase samples contain only a limited number of spectral components and both the amplitude and wave vectors of these components are random. Therefore, the required number of spectral components is significantly lower than the number of spatial points in the phase screen.

On similar lines as in conventional FFT based models, let the two-dimensional complex random phase $\theta_{SS}(r)$ be expressed as sum of harmonics,

$$\theta_{SS}(r) = \sum_{n=1}^{N} a_n \exp(i\kappa_n \cdot r) = \theta_1(r) + i\theta_2(r), \qquad (8.64)$$

where $a_n$ denotes the random complex spectral amplitudes, $\kappa_n = (\kappa_{xn}, \kappa_{yn})$ denotes the wave vectors of individual spectral components, and $N$ represents the total number of spectral components used to represent phase [78]. In order to represent statistically homogeneous, zero average random phase, the first and second moments of the random spectral amplitudes $a_n$ are taken as:

$$\langle a_n \rangle = 0, \ \langle a_n a_m \rangle = 0 \text{ and } \langle a_n a_m^* \rangle = s_n \delta_{nm}. \qquad (8.65)$$

The wave vectors $\kappa_n$ are assumed to be random vectors with probability distributions,

$$P(\kappa_n \in (\kappa + d\kappa)) = p_n(\kappa)d\kappa, \qquad (8.66)$$

and the ensemble averaged structure function for the phase screens of SS model is simply,

$$D_\theta^{SS}(r) = \langle [\theta_{SS}(r_1) - \theta_{SS}(r_2)]^2 \rangle_{\{a_n,\kappa_n\}}, \qquad (8.67)$$

$$D_\theta^{SS}(r) = \left\langle \sum_{n=1}^{N} s_n[1 - \cos(\kappa_n \cdot r)] \right\rangle_{\{\kappa_n\}}, \qquad (8.68)$$

$$D_\theta^{SS}(r) = \iint d^2\kappa \sum_{n=1}^{N} s_n p_n(\kappa)[1 - \cos(\kappa_n \cdot r)]. \qquad (8.69)$$

This would match the target structure function given in equation (8.41) if the following condition is satisfied:

$$\sum_{n=1}^{N} s_n p_n(\kappa) = 2\Phi_\theta(\kappa). \qquad (8.70)$$

The above equation decides the constraints on weights $s_n$ and the probability distribution $p_n(\kappa)$. The conventional FFT phase screen model can still be obtained from the SS model in the form of equation (8.64) when $p_n(\kappa)$ are delta functions supported at the nodes of the rectangular grid [80]. For isotropic spectra $\Phi_\theta(\kappa)$, the wave vector's probability density functions $p_n(\kappa)$ can be chosen in polar coordinates:

$$p_n(\boldsymbol{\kappa})d^2\kappa = p_n(\kappa, \phi)\kappa d\kappa d\phi = p_n(\kappa)d\kappa \frac{d\phi}{2\pi}. \tag{8.71}$$

The above equation assumes that the wave vector's directions are uniformly distributed on the interval $[-\pi, +\pi]$ in line with the statistical isotropy of the phase spectrum. There can be a number of different ways to satisfy equation (8.70), and each one of them will result in different Monte Carlo models for generation of the phase samples with the given structure function. Charnotskii has discussed two types of models (a) the partitioned model (PM) and (b) the overlap model (OM) which are mathematically quite different from each other [81]. Here, we will briefly discuss the partitioned model. For the PM model, a non-overlapping partition of the desired range of wave-numbers $\kappa_{min} < \kappa < \kappa_{max}$ into $N$ sub-intervals is created so that

$$\kappa_{min} < \kappa_1 < \kappa_2 < \cdots < \kappa_{N-1} < \kappa_N = \kappa_{max}, \tag{8.72}$$

and the probability distribution $p_n(\kappa)$ is supported only on the $n$th sub-interval $\kappa_{n-1} < \kappa < \kappa_n$. Hence,

$$s_n = 2\pi \int_{\kappa_{n-1}}^{\kappa_n} \kappa d\kappa \, \Phi_\theta(\kappa), \tag{8.73}$$

and,

$$p_n(\kappa) = \begin{cases} \kappa \, \Phi_\theta(\kappa)\left[\int_{\kappa_{n-1}}^{\kappa_n} \kappa d\kappa \, \Phi_\theta(\kappa)\right]^{-1}, & \kappa_{n-1} < \kappa < \kappa_n \\ 0, & \kappa < \kappa_{n-1} \text{ or } \kappa > \kappa_n. \end{cases} \tag{8.74}$$

Log-uniform partition of wave-numbers has been identified as the most efficient for turbulence phase screen modeling [81]. Thus, the partition wave-numbers in equation(8.74) are expressed as,

$$\kappa_n = \kappa_{min} \exp\left[\frac{n}{N} \ln\left(\frac{\kappa_{max}}{\kappa_{min}}\right)\right]. \tag{8.75}$$

Generating practical realization of random wave-numbers with probability distributions $p_n(\kappa)$ has been shown to be straightforward for pure power-law spectra [79, 81]. For more complicated spectra such as the von Karman spectrum, such realizations can be computationally expensive. A major disadvantage of the sparse spectrum method is that it requires direct series summation of terms and one cannot make use of the FFT algorithm to calculate the phase realizations. However, as the method makes use of a moderate number of spectral components in the sparse spectrum series representation, the overall computational time for several mega-pixel-size phase screens is not too large [79]. Also, the phase samples produced using this method are unbiased and non-periodic.

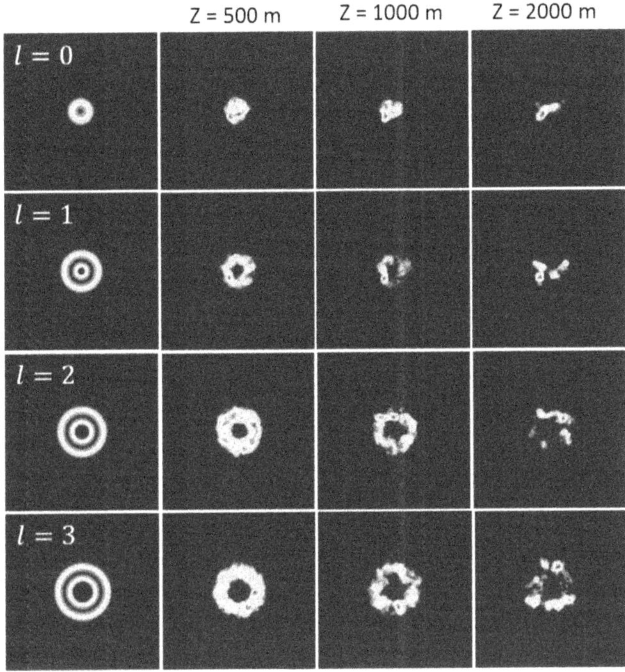


**Figure 8.11.** Illustration of the split-step method showing one realization of propagation of OAM states $l = 0$, 1, 2, 3 through atmospheric turbulence. The beam waists parameter used for all the beams is the same and given by $w = 5$ cm and the wavelength of the laser is $\lambda = 1.55$ $\mu$m. The turbulence strength is given by $C_n^2 = 10^{-14}$ m$^{-2/3}$ and the von Karman spectrum has been used. The beam intensity profiles at distances $z = 500$ m, 1000 m and 2000 m are shown. All the patterns are normalized individually for maximum dynamic range of display in individual sub-figures.

## 8.4 Illustration of propagation of OAM states through turbulence

As an illustration of the computational methods described in this section we now show one realization of propagation of OAM states with $l = 0$, 1, 2, 3 in figure 8.11 OAM states have received much attention as communication channels for free space communication systems [82, 83]. Rigorous turbulence modeling may also be needed for understanding degradation of entanglement when one or both the entangled photons travel through turbulence [84, 85]. As the interest in propagating individual OAM states or their superposition through realistic atmospheric conditions is growing, it is important for a number of researchers to quickly model such turbulence propagation in order to test the performance of their system designs of interest. In the illustration shown below, the inner and outer turbulence scale values of $l_o = 1$ cm and $L_o = 3$ m were used and the turbulence strength was assumed to be $C_n^2 = 10^{-14}$ m$^{-2/3}$ along with von Karman spectrum. The sampling grid is $512 \times 512$ pixels with a sampling interval of 1 mm. The laser wavelength is $\lambda = 1.55$ $\mu$m. The initial intensity profile of the beam is shown in the left column of figure 8.11, followed by intensity profiles at a propagation distance of 500 m, 1000 m and 2000 m respectively. The split-step method used 20 random phase screens placed over a

2 km distance. The illustration shows that the intensity profile of the OAM states is fairly undisturbed till 500 m but is highly degraded over a 2 km propagation distance. Detection and separation of OAM modes as required in applications such as free space communication can therefore pose challenges over long-range turbulence paths. In the next chapter we will provide discussion on how collinear propagation of OAM states (particularly the $l = 0, 1$ states) can provide robust laser beam designs that are able to maintain their central intensity lobes on long-range turbulence propagation. The annotated software codes as described in appendix A can be implemented quickly to get an idea of beam intensity profile on propagation through turbulence. It may be noted that while beam intensity profiles are shown in figure 8.11, the phase of the propagated fields is available simultaneously. Such computed phase profiles can, for example, be utilized for estimating the nature of phase correction that may be required in an adaptive optical system.

## 8.5 Beam quality parameters

The amplitude and phase profile of the laser beam gets distorted on propagation through the random phase screen. A few parameters such as the amount of beam spread and beam wander, scintillation index, signal-to-noise ratio, bit-error-rate, average intensity, $M^2$ factor and the spatial coherence of the beam can be used to quantify the quality of the distorted beam. Two of these methods are described in the following discussion.

### 8.5.1 Scintillation index

Consider a laser beam propagating in atmosphere and falling onto a receiver situated in any $z$ plane. If we attempt to measure its intensity (or irradiance) at the receiver, we will find that its instantaneous intensity $I(x, y, z)$ will fluctuate in time about its average value $\langle I(x, y, z) \rangle$. These fluctuations in the beam's irradiance are known as scintillations [86]. The beam scintillations arise due to two reasons: (a) the beam intensity profile on passing through the random phase screen breaks up into randomly shaped speckles and (b) the beam experiences random deflections resulting in wandering of its centroid position. Beam scintillation leads to power losses at the receiver resulting in significantly low signal-to-noise ratio and eventually signal fading. Therefore, it is desirable to be able to predict the magnitude of intensity scintillation as an important design consideration.

The statistics concerning irradiance fluctuations and phase fluctuations are obtained from the fourth moment of the optical field. The scintillation index (or SI) is defined at any given transverse position $(x, y)$ as:

$$\sigma_I^2(x, y, z) = \frac{\langle I(x, y, z)^2 \rangle}{\langle I(x, y, z) \rangle^2} - 1, \tag{8.76}$$

where the average intensity $\langle I(x, y, z) \rangle$ is obtained by summing over independently obtained realizations of intensity values and then dividing the result by the total number of realizations. Similarly for $\langle I(x, y, z)^2 \rangle$, the square of the intensity values of independent realizations is first added together and then averaged. The usual practise is to look at the on-axis SI, i.e., when $x = y = 0$. For plane waves, the irradiance $I(x, y, z)$ at a point detector centered at $(x = 0, y = 0)$ is enough to calculate the on-axis SI. However, in real applications, the detectors have some aperture size which gives rise to the aperture averaging effects, thus reducing the measured intensity fluctuations [87–89]. In this case, the nature of the observed intensity fluctuations depend on the size of the receiver aperture. At a given instant of time, the observable quantity then becomes,

$$S(z) = \int_A I(x, y, z)dxdy, \tag{8.77}$$

where $A$ is the surface area of the detector aperture and $I(x, y, z)$ is the instantaneous irradiance. The SI is now computed using $S(z)$,

$$\sigma_S^2 = \frac{\langle S^2 \rangle}{\langle S \rangle^2} - 1. \tag{8.78}$$

Thus, the detector aperture averages the fluctuations of the received beam over the aperture area, leading to reduced signal fluctuations compared to a point detector. The on-axis scintillation index describes the intensity fluctuations at a single on-axis point in the receiver plane. However, intensity fluctuations at one point of the beam are correlated with those at another point. This spatial structure of scintillation is studied using the co-variance function of irradiance [1]. A characteristic correlation width can be defined as either the first zero or the $1/e^2$ point of the normalized co-variance function. Aperture sizes of the order of correlation width or smaller will act like a 'point' receiver whereas apertures greater than correlation width will cause reduction in observed scintillation due to aperture averaging. Correlation length is related to the size of the first Fresnel zone which is given by $\sqrt{L/k}$. In practice, a detector aperture of diameter less than the Fresnel length is treated as a point detector for measuring scintillation. Therefore, in simulation studies, the usual practice is to choose a detector aperture whose radius should be below $\sqrt{\lambda L/2\pi}$.

The theoretical aspects of beam scintillation have been studied over several decades, starting with Tatarskii [3, 4]. The theory of optical scintillation for plane, spherical and Gaussian beams have been established in weak turbulence conditions and the review of the same can be found in [86].

### 8.5.2 Signal-to-noise ratio for instantaneous beam profile

The signal-to-noise ratio (SNR) of the beam intensity is defined as the mean intensity divided by the standard deviation of intensity values within a given detector area,

$$SNR = \frac{\langle I \rangle}{\sigma_I}. \tag{8.79}$$

The SNR is calculated over instantaneous intensity values which is then averaged over the total number of realizations. It is important to note here that the on-axis SI conventionally makes use of the long-time averaged intensity values while SNR uses instantaneous intensity values. In the next chapter on robust beam engineering, we will find that the instantaneous SNR is very useful in understanding the evolution of structured light beams through turbulence.

We conclude by noting that well-annotated computer simulation codes for the main numerical methods discussed in this chapter are provided in appendix A. They can be handy for beginning researchers who wish to perform simulation tests for propagation of beams with arbitrary amplitude-phase and polarization profile through turbulence.

# References


[1] Andrews L C and Phillips R L 2005 *Laser Beam Propagation Through Random Media* (Bellingham, WA: SPIE Optical Engineering Press)

[2] Levy M 2000 *Parabolic Equation Methods for Electromagnetic Wave Propagation (IEE Electromagnetic Wave Series 45)* (London: Institution of Engineering and Technology)

[3] Tatarskii V I 1961 *Wave Propagation in a Turbulent Medium* (New York: McGraw-Hill) (volume translated by Richard A Silverman)

[4] Tatarskii V I 1971 *The Effects of the Turbulent Atmosphere on Wave Propagation* (Washington, DC: National Oceanic and Atmospheric Administration, U.S. Department of Commerce and the National Science Foundation)

[5] Wheelon A D 2001 *Electromagnetic Scintillation* vol 1 (Cambridge: Cambridge University Press)

[6] Fante R L 1975 Electromagnetic beam propagation in turbulent media *Proc. IEEE* **63** 1669–92

[7] Strohbehn J W 1978 *Laser Beam Propagation in the Atmosphere (Topics in Applied Physics vol 25)* (Berlin: Springer)

[8] Strohbehn J W 1968 Line-of-sight wave propagation through the turbulent atmosphere *Proc. IEEE* **56** 1301–8

[9] Belmonte A 2000 Feasibility study for the simulation of beam propagation: consideration of coherent lidar performance *Appl. Opt.* **39** 5426–45

[10] Prokhorov A M, Bunkin F V, Gochelashvily K S and Shishov V I 1975 Laser irradiance propagation in turbulent media *Proc. IEEE* **63** 790–811

[11] Frehlich R G 1987 Intensity covariance of a point source in a random medium with a Kolmogorov spectrum and an inner scale of turbulence *J. Opt. Soc. Am.* A **4** 360–6

[12] Wang G-Y and Dashen R 1993 Intensity moments for waves in random media: three-order standard asymptotic calculation *J. Opt. Soc. Am.* A **10** 1226–32

[13] Dashen R and Wang G-Y 1993 Intensity fluctuation for waves behind a phase screen: a new asymptotic scheme *J. Opt. Soc. Am.* A **10** 1219–25

[14] Feizulin Z I and Kravtsov Yu A 1967 Broadening of a laser beam in a turbulent medium *Radiophys. Quantum Electron.* **10** 33–5

[15] Lutomirski R F and Yura H T 1971 Propagation of a finite optical beam in an inhomogeneous medium *Appl. Opt.* **10** 1652–8

[16] Charnotskii M 2015 Extended huygen's fresnel principle and optical waves propagation in turbulence: discussion *J. Opt. Soc. Am.* A **32** 1357–65

[17] Tyson R K 2016 *Principles of Adaptive Optics* 4th edn (Boca Raton, FL: CRC Press)

[18] Roggemann M C and Welsh B M 1996 *Imaging Through Turbulence* (Boca Raton, FL: CRC Press)

[19] Marchuk G I 1980 *The Monte Carlo Methods in Atmospheric Optics (Springer Series in Optical Sciences)* (Berlin: Springer)

[20] Uscinski B J 1993 Multi-phase-screen analysis *Wave Propagation in Random Media (Scintillation)* ed V I Tatarskii, A Ishimaru and V U Zavorotny (Bellingham, WA: SPIE Optical Engineering Press) p 346

[21] Macaskill C and Ewart T E 1984 Computer simulation of two-dimensional random wave propagation *IMA J. Appl. Math.* **33** 1–15

[22] Spivack M and Uscinski B J 1989 The split-step solution in random wave propagation *J. Comput. Appl. Math.* **27** 349–61

[23] Schmidt J D 2010 *Numerical Simulation of Optical Wave Propagation with Examples in MATLAB* (Bellingham, WA: SPIE Optical Engineering Press)

[24] Martin J M and Flatte S M 1988 Intensity images and statistics from numerical simulation of wave propagation in 3d random media *Appl. Opt.* **27** 2111–26

[25] Coles Wm A, Filice J P, Frehlich R G and Yadlowsky M 1995 Simulation of wave propagation in three-dimensional random media *Appl. Opt.* **34** 2089–101

[26] Johansson E M and Gavel D T 1994 Simulation of stellar speckle imaging *Amplitude and Intensity Spatial Interferometry II* vol 2200, ed J B Breckinridge (Bellingham, WA: SPIE Optical Engineering Press) pp 372–83

[27] Knepp D L 1983 Multiple phase-screen calculation of the temporal behavior of stochastic waves *Proc. IEEE* **71** 722–37

[28] Martin J M and Flatté S M 1990 Simulation of point-source scintillation through three-dimensional random media *J. Opt. Soc. Am.* A **7** 838–47

[29] Voelz D G and Xiao X 2009 Metric for optimizing spatially partially coherent beams for propagation through turbulence *Opt. Eng.* **48** 036001

[30] Dockery G D 1988 Modeling electromagnetic wave propagation in the troposphere using the parabolic equation *IEEE Trans. Antennas Propag.* **36** 1464–70

[31] Kuttler J R and Dockery G D 1991 Theoretical description of the parabolic approximation/ Fourier split-step method of representing electromagnetic propagation in the troposphere *Radio Sci.* **26** 381–93

[32] Morris J R, Fleck J A and Feit M D 1976 Time-dependent propagation of high energy laser beams through the atmosphere *Appl. Phys.* **10** 129–60

[33] Weideman J A C and Herbst B M 1986 Split-step methods for the solution of the nonlinear Schrödinger equation *SIAM J. Numer. Anal.* **23** 485–507

[34] Leontovich M A and Fock V A 1946 Solution of the problem of propagation of electro-magnetic waves along the Earth's surface by method of parabolic equations *J. Phys. USSR* **10** 13–23

[35] Hardin R H and Tappert F D 1973 Applications of the split-step Fourier method to the numerical solution of nonlinear and variable coefficient wave equations *SIAM Rev.* **15** 423

[36] Flatte S M and Tappert F D 1975 Calculation of the effect of internal waves on oceanic sound transmission *J. Acoust. Soc. Am.* **58** 1151–9

[37] DiNapoli F R and Deavenport R L 1977 *Numerical Methods of Underwater Acoustic Propagation* (Berlin: Springer)

[38] Tappert F D 1974 Parabolic equation method in underwater acoustics *J. Acoust. Soc. Am.* **55**

[39] Goodman J 2004 *Introduction to Fourier Optics* 3rd edn (New Delhi: Viva Books)

[40] Voelz D G and Roggemann M C 2009 Digital simulation of scalar optical diffraction: revisiting chirp function sampling criteria and consequences *Appl. Opt.* **48** 6132–42

[41] Liu J-P 2012 Controlling the aliasing by zero-padding in the digital calculation of the scalar diffraction *J. Opt. Soc. Am.* A **29** 1956–64

[42] Mas D, Garcia J, Ferreira C, Bernardo L M and Marinho F 1999 Fast algorithms for free-space diffraction patterns calculation *Opt. Commun.* **164** 233–45

[43] Lochab P P, Senthilkumaran P and Khare K 2019 Propapation of converging polarization singular beams through atmospheric turbulence *Appl. Opt.* **58** 6335–45

[44] Matsushima K and Shimobaba T 2009 Band-limited angular spectrum method for numerical simulation of free-space propagation in far and near fields *Opt. Express* **17** 19662–73

[45] Yura H T 1979 Signal-to-noise ratio of heterodyne lidar systems in the presence of atmospheric turbulence *Opt. Acta: Int. J. Opt.* **26** 627–44

[46] Xiao X and Voelz D 2009 On-axis probability density function and fade behavior of partially coherent beams propagating through turbulence *Appl. Opt.* **48** 167–75

[47] Bramley E N, Appleton and Victor E 1954 The diffraction of waves by an irregular refracting medium *Proc. R. Soc. Lond. Ser.* A **225** 515–8

[48] Ratcliffe J A 1956 Some aspects of diffraction theory and their application to the ionosphere *Rep. Prog. Phys.* **19** 188–267

[49] Booker H G, Ferguson J A and Vats H O 1985 Comparison between the extended-medium and the phase-screen scintillation theories *J. Atmos. Terr. Phys.* **47** 381–99

[50] Uscinski B J 1985 Analytical solution of the fourth-moment equation and interpretation as a set of phase screens *J. Opt. Soc. Am.* A **2** 2077–91

[51] Welsh B M 1997 Fourier-series-based atmospheric phase screen generator for simulating anisoplanatic geometries and temporal evolution *Propagation and Imaging through the Atmosphere* vol 3125, ed L R Bissonnette and C Dainty (Bellingham, WA: SPIE Optical Engineering Press) pp 327–38

[52] Roddier N A 1990 Atmospheric wavefront simulation using Zernike polynomials *Opt. Eng.* **29** 1174–80

[53] Mitra S K 2001 *Digital Signal Processing: a Computer-Based Approach* (New York: McGraw-Hill/Irwin)

[54] Herman B J and Strugala L A 1990 Method for inclusion of low-frequency contributions in numerical representation of atmospheric turbulence *Propagation of High-Energy Laser Beams Through the Earth's Atmosphere* vol 1221, ed P B Ulrich and L E Wilson (Bellingham, WA: SPIE Optical Engineering Press) pp 183–92

[55] Noll R J 1976 Zernike polynomials and atmospheric turbulence *J. Opt. Soc. Am.* **66** 207–11

[56] Xiang J 2012 Accurate compensation of the low-frequency components for the FFt-based turbulent phase screen *Opt. Express* **20** 681–7

[57] Vorontsov A M, Paramonov P V, Valley M T and Vorontsov M A 2008 Generation of infinitely long phase screens for modeling of optical wave propagation in atmospheric turbulence *Waves Random Complex Media* **18** 91–108

[58] Lane R G, Glindemann A and Dainty J C 1992 Simulation of a Kolmogorov phase screen *Waves Random Media* **2** 209–24

[59] Frehlich R 2000 Simulation of laser propagation in a turbulent atmosphere *Appl. Opt.* **39** 393–7

[60] Roggemann M C 2014 Simulating non-Kolomogorov phase screens with finite inner and outer scales *Imaging and Applied Optics 2014* (Washington, DC: Optical Society of America) p PM4E.1

[61] Banakh V A, Krekov G M, Mironov V L, Khmelevtsov S S and Tsvik R S 1974 Focused-laser-beam scintillations in the turbulent atmosphere *J. Opt. Soc. Am.* **64** 516–8

[62] Sedmak G 2004 Implementation of fast-fourier-transform-based simulations of extra-large atmospheric phase and scintillation screens *Appl. Opt.* **43** 4527–38

[63] Carbillet M and Riccardi A 2010 Numerical modeling of atmospherically perturbed phase screens: new solutions for classical fast Fourier transform and Zernike methods *Appl. Opt.* **49** G47–52

[64] Liu T, Zhang J, Lei Y, Zhang R and Sun J 2019 Optimal subharmonics selection for atmosphere turbulence phase screen simulation using the subharmonic method *J. Mod. Opt.* **66** 986–91

[65] Glindemann A, Lane R G and Dainty J C 1992 Simulation of time-evolving speckle patterns using Kolmogorov statistics *J. Mod. Opt.* **40** 2381–8

[66] Harding C M, Johnston R A and Lane R G 1999 Fast simulation of a Kolmogorov phase screen *Appl. Opt.* **38** 2161–70

[67] Johnston R A and Lane R G 2000 Modeling scintillation from an aperiodic Kolmogorov phase screen *Appl. Opt.* **39** 4761–9

[68] Roggemann M C, Welsh B M, Montera D and Rhoadarmer T A 1995 Method for simulating atmospheric turbulence phase effects for multiple time slices and anisoplanatic conditions *Appl. Opt.* **34** 4037–51

[69] Beghi A, Cenedese A and Masiero A 2008 Stochastic realization approach to the efficient simulation of phase screens *J. Opt. Soc. Am.* A **25** 515–25

[70] Sriram V and Kearney D 2007 An ultra fast Kolmogorov phase screen generator suitable for parallel implementation *Opt. Express* **15** 13709–14

[71] Dios F, Recolons J, Rodríguez A and Batet O 2008 Temporal analysis of laser beam propagation in the atmosphere using computer-generated long phase screens *Opt. Express* **16** 2206–20

[72] Kouznetsov D, Voitsekhovich V V and Ortega-Martinez R 1997 Simulations of turbulence-induced phase and log-amplitude distortions *Appl. Opt.* **36** 464–9

[73] Eckert R J and Goda M E 2006 Polar phase screens: a comparative analysis with other methods of random phase screen generation *Atmospheric Optical Modeling, Measurement, and Simulation II* vol 6303, ed S M Hammel and A Kohnle (Bellingham, WA: SPIE Optical Engineering Press) pp 1–14

[74] Jakobsson H 1996 Simulations of time series of atmospherically distorted wave fronts *Appl. Opt.* **35** 1561–5

[75] Assémat F, Wilson R W and Gendron E 2006 Method for simulating infinitely long and non stationary phase screens with optimized memory storage *Opt. Express* **14** 988–99

[76] Goldring T and Carlson L 1989 Analysis and implementation of non-Kolmogorov phase screens appropriate to structured environments *Nonlinear Optical Beam Manipulation and High Energy Beam Propagation Through the Atmosphere* vol 1060, ed R A Fisher and L E Wilson (Bellingham, WA: SPIE Optical Engineering Press) pp 244–64

[77] Paulson D A, Wu C and Davis C C 2019 Randomized spectral sampling for efficient simulation of laser propagation through optical turbulence *J. Opt. Soc. Am.* B **36** 3249–62

[78] Charnotskii M 2020 Comparison of four techniques for turbulent phase screens simulation *J. Opt. Soc. Am.* A **37** 738–47

[79] American Society of Mechanical Engineers—Ocean, Offshore and Arctic Engineering Division 2011 Sparse spectrum model of the sea surface *30th Int. Conf. on Ocean, Offshore and Arctic Engineering* vol 6: Ocean Engineering (New York: ASME)

[80] Charnotskii M 2013 Sparse spectrum model for a turbulent phase *J. Opt. Soc. Am.* A **30** 479–88

[81] Charnotskii M 2013 Statistics of the sparse spectrum turbulent phase *J. Opt. Soc. Am.* A **30** 2455–65

[82] Malik M, O'Sullivan M, Rodenburg B, Mirhosseini M, Leach J, Lavery M P J, Padgett M J and Boyd R W 2012 Influence of atmospheric turbulence on optical communications using orbital angular momentum for encoding *Opt. Express* **20** 13195–200

[83] Krenn M, Fickler R, Fink M, Handsteiner J, Malik M, Scheidl T, Ursin R and Zeilinger A 2014 Communication with spatially modulated light through turbulent air across Vienna *New J. Phys.* **16** 113028

[84] Roux F S 2011 Infinitesimal-propagation equation for decoherence of an orbital-angular-momentum-entangled biphoton state in atmospheric turbulence *Phys. Rev.* A **83** 053822

[85] Zhang Y *et al* 2016 Experimentally observed decay of high-dimensional entanglement through turbulence *Phys. Rev.* A **94** 032310

[86] Andrews L C, Phillips R L and Young C Y 2001 *Laser Beam Scintillation with Applications* (Bellingham, WA: SPIE Optical Engineering Press)

[87] Fried D L 1967 Aperture averaging of scintillation *J. Opt. Soc. Am.* **57** 169–75

[88] Andrews L C 1992 Aperture-averaging factor for optical scintillations of plane and spherical waves in the atmosphere *J. Opt. Soc. Am.* A **9** 597–600

[89] Wang S J, Baykal Y and Plonus M A 1983 Receiver-aperture averaging effects for the intensity fluctuation of a beam wave in the turbulent atmosphere *J. Opt. Soc. Am.* **73** 831–7

**IOP** Publishing

Orbital Angular Momentum States of Light
Propagation through atmospheric turbulence
Kedar Khare, Priyanka Lochab and Paramasivam Senthilkumaran

# Chapter 9

# Robust laser beam engineering using complementary diffraction

This chapter describes a robust beam engineering principle that has resulted out of the recent work of the authors of this book. The intensity profile of optical beams on encountering a spatially varying refractive index gets distorted beyond the nominal beam spreading due to diffraction effects. Refractive index variations of the order of $10^{-5}$–$10^{-6}$ are sufficient to make the beam profile speckled. This poses a serious problem for multiple applications like free space optical communication, underwater communication, laser guided defense systems, LIght Detection And Ranging (LIDAR), etc, which depend on laser beam propagation through a random medium like atmosphere, ocean or biological tissues. The performance of these systems is greatly limited because the random time-varying fluctuations in the beam intensity compromise with the amount of energy that can be delivered by the beam. This beam degradation essentially leads to signal fading and broken optical links. This long standing problem of sending information or energy through a random medium by optical beams has attracted lot of attention from the scientific community. Various types of beam designs dealing with the beam properties like beam profile, polarization, coherence, etc, have been proposed and tested [1–8]. However, it is not easy to provide specific guidelines for generating robust beams for propagation through a random medium due to the stochastic nature of the problem. In this chapter, we take a step in this direction and provide a novel idea of designing robust vector beams using the speckle diversity principle. The basic aim is to evolve an optimal beam engineering approach using the interplay between polarization and OAM states of light that is inherently best suited for sending light beams through a turbulent medium.

In this chapter we will first discuss a peculiar property associated with diffraction patterns observed when a general aperture is illuminated using the $l = 0$ and $l = 1$ OAM states and use it further to develop a robust beam engineering principle that explicitly uses the polarization singular beams explained in chapter 6.

Our simulations and experiments suggest that this principle may be used to design beams that maintain their intensity profile even after propagation through random phase fluctuations as encountered in atmosphere. For simplicity the atmosphere will be modeled first using a single phase screen. Next we will validate the robust beam engineering principle for long range turbulence propagation using the split-step method.

## 9.1 Complementary diffraction due to (0,1) OAM states

In chapter 2 we developed the notion of the spiral phase quadrature transform as a two-dimensional analog of the Hilbert transform. It was shown that two functions $g_1(x, y)$ and $g_2(x, y)$ connected by the spiral phase filter $\exp(i\phi)$ ($\phi$ is the polar angle in two-dimensional Fourier space) approximately had sine-cosine like behavior in the sense of Mandel's theorem. The simplest way to realize the above result in an optics laboratory is to illuminate an arbitrary amplitude-phase aperture with $l = 0$ and $l = 1$ OAM states that distinctly differ in their phase profile which is given by $\exp(il\phi)$. The result that the spiral phase filter is an analogue of the Hilbert transform implies that the Fraunhofer diffraction patterns observed due to these two illuminations for a given aperture must show sine–cosine like diversity. We illustrate this effect first with diffraction patterns associated with simple apertures. Figure 9.1 shows the Fraunhofer diffraction intensity patterns corresponding to circular, square and rectangular (slit-like) apertures when illuminated by the $l = 0$ and $l = 1$ OAM states. In particular we observe the maxima and minima in the two diffraction intensity patterns due to the two OAM states getting exchanged. It should be observed that this complementarity is local just like in a sine–cosine pair. As we will see later in this section, the complementary diffraction is simply a property of the

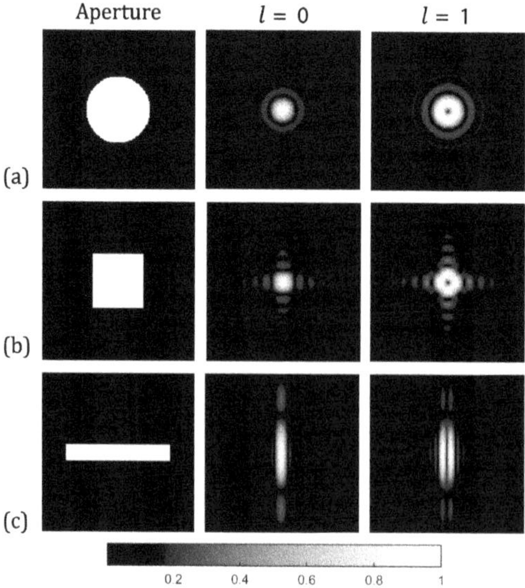


**Figure 9.1.** Diffraction patterns corresponding to the $l = 0$ and $l = 1$ OAM states are shown for (a) circular, (b) square and (c) rectangular apertures respectively.

spiral phase quadrature transform and should, therefore, also be observed for general random phase apertures as well. One case of interest to multiple applications is the propagation of optical beams through atmospheric turbulence. A random medium like atmosphere at a given instant can be approximately described by a suitable phase screen in the lens aperture. Therefore, we next simulate the diffraction of these two OAM states through a random phase screen. In order to generate the random phase screen, we first take a random phase function in the lens aperture with phase distributed uniformly in $[0,\ 2\pi]$. This random phase function is then convolved with an averaging Gaussian filter to induce correlation over phase fluctuation length scales. In later sections, we will model the atmosphere with random phase screens generated using the FFT method as explained in previous chapters. For now, the present method of generating a random phase screen is sufficient for illustration of complementarity. Figure 9.2 shows the diffraction patterns corresponding to $l = 0$ and $l = 1$ OAM states when a random phase screen was introduced in the lens aperture plane. Individually the diffraction spot has degraded to a speckled appearance due to the random phase in the aperture. The complementary nature of these diffraction patterns can however be easily observed for the case of random phase screen apertures as well. In order to quantify the complementarity, we

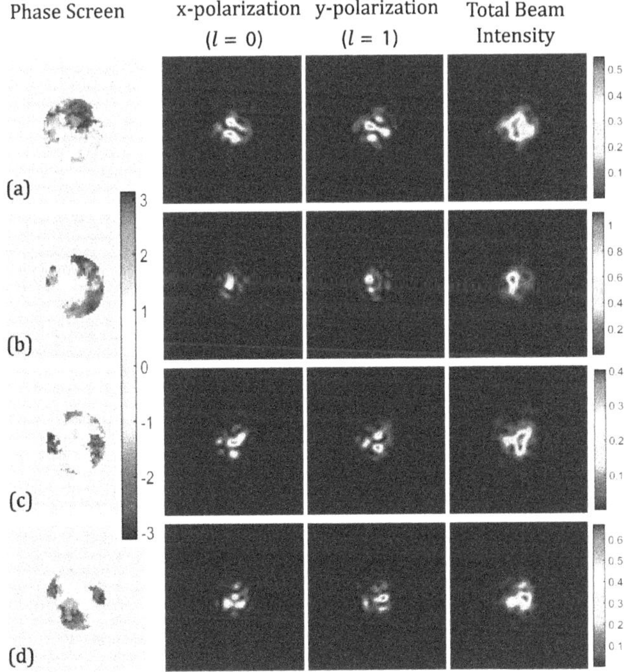


**Figure 9.2.** The simulated intensities corresponding to the $l = 0, 1$ illumination are shown for four different random phase screens which are placed in the lens aperture.

calculate the correlation coefficient $(\rho_{I_1,I_2})$ between the intensities $I_1$ and $I_2$ corresponding to the OAM states $l = 0$ and $l = 1$. The correlation coefficient is defined as:

$$\rho_{I_1,I_2} = \frac{\langle [(I_1 - \mu_1)(I_2 - \mu_2)] \rangle}{\sigma_{I_1} \sigma_{I_2}}, \tag{9.1}$$

where $\langle\ \rangle$ denote the ensemble average and $\mu$ and $\sigma$ are the mean and the standard deviation respectively. The correlation coefficient is a measure of the strength of linear association between two variables. It can take any value in the range $[-1, +1]$ with $+1$ (or $-1$) indicating perfect positive (or negative) correlation. The positive correlation means that any increase (or decrease) in the value of one variable is accompanied by increase (or decrease) in the value of the second variable. A negative correlation on the other hand means that one variable increases as the second variable decreases and vice versa. A value of zero would correspond to no correlation between the two variables. In the present case, the diffraction patterns of the $l = 0$ and $l = 1$ OAM states would be complementary if the corresponding correlation coefficient has a sufficiently large negative value. Based on 50 realizations of the random phase screen, the correlation coefficient for the intensity images for $l = 0$ and $l = 1$ OAM states is observed to be equal to $\rho_{I_1,I_2} = -0.95 \pm 0.04$. This value of $\rho_{I_1,I_2}$ has been evaluated over a region equal to the free space diffraction-limited spot size of the $l = 0$ OAM mode. Thus, the two diffraction intensity patterns are in fact highly negatively correlated. This high negative correlation suggests that if the diffraction patterns corresponding to $l = 0$ and $l = 1$ OAM states are added together, the resulting total intensity profile should be more uniform than the individual diffraction pattern intensities. The last column in figure 9.2 shows this combined (or total) beam intensity profile. It can be observed that the combined intensity profile has a well-defined central lobe even when different random phase screens are used. Therefore, when the $l = 0$ and $l = 1$ OAM states are propagated collinearly through a given random phase screen, the complementary diffraction intensity patterns when added together can yield an improved intensity profile. In practice the random phase screens may represent different realizations of fluctuating atmosphere that are time-varying. The complementary diffraction effect will still remain valid thereby maintaining beam quality through time-varying random phase fluctuations.

It is important to note that incoherent addition of two diffraction intensity patterns is required for this robust beam engineering approach. The incoherent addition can be achieved for example by engineering the beam such that the two OAM states are embedded in two orthogonal polarizations. Another possible mechanism is to combine two independent laser beams made to operate in the $l = 0, 1$ spatial modes. In the present work, we will take the polarization approach for convenience. This approach will also allow us naturally to describe the engineered beams in terms of the polarization singularities formalism. Choosing the $l = 0$ OAM state to be in $x$ polarization and $l = 1$ OAM state in the $y$ polarization, the engineered vector beam propagating in $z$ direction could be described in the $z = 0$ plane as,

$$\vec{E}(x, y) = \frac{1}{\sqrt{2}}[\hat{x} + e^{i\theta}\hat{y}]\psi(x, y), \tag{9.2}$$

where the function $\psi(x, y)$ represents the normalized beam profile function for example, the Gaussian or LG(0,0) mode or any other native beam profile. A time dependence of form $\exp(-i\omega t)$ is assumed. Here, $\hat{x}$ and $\hat{y}$ represent unit vectors in the $x$ and $y$ directions respectively and $\theta$ is the polar angle. To experimentally realize this beam, one may use the arrangement involving a spatial light modulator as shown in figure 9.3. A 45-degree linearly polarized laser beam in an $l = 0$ OAM state is made incident on a reflective phase spatial light modulator (SLM). The liquid crystal SLMs are usually sensitive to one polarization and act like a plane mirror for the orthogonal polarization. The vortex phase $\theta$ is displayed on the SLM. On reflection from the SLM, both the polarization components get reflected collinearly with one of the polarizations in $l = 1$ OAM state to lead to the beam described in equation (9.2). The polarization structure of the reflected beam at the SLM plane and after propagation from SLM plane is shown in figure 9.4. From the discussion in chapter 6 it is clear that the vector beam above contains C-point polarization singularity structures (lemon, star separated by L-line) embedded in it after it has propagated from the SLM plane. The polarization map has been generated after numerically propagating the scalar components in equation (9.2). If the individual scalar components are passed through the random phase screen shown in figure 9.3, the diffraction pattern may be observed

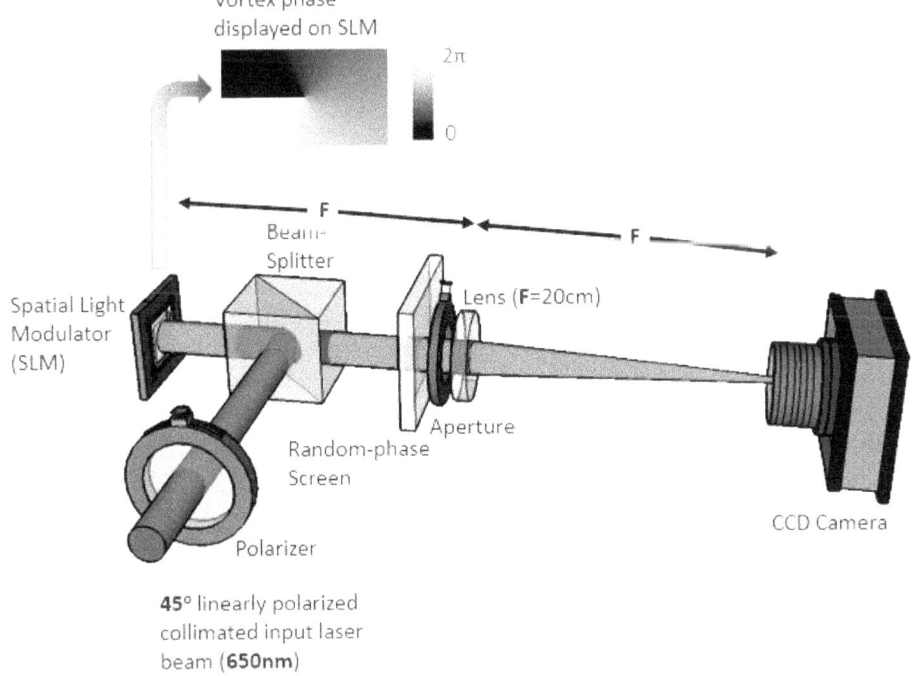


**Figure 9.3.** Experimental set-up for a generation vector beam made as a combination of $l = 0$, 1 OAM states in two orthogonal polarizations. Reprinted with permission from [9] © The Optical Society.

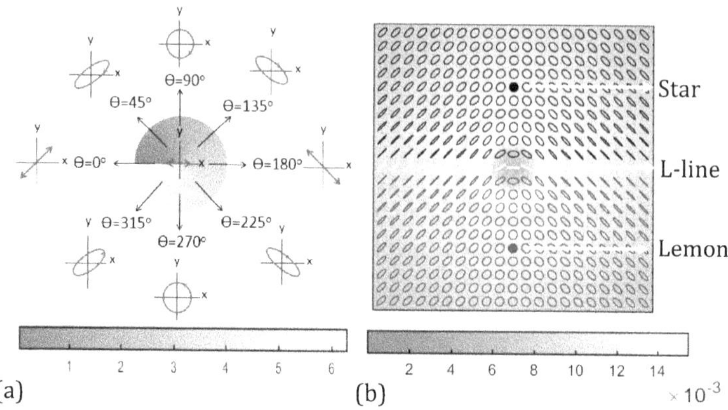



**Figure 9.4.** The state of polarization of the engineered vector beam is shown: (a) at the SLM plane and (b) after free space propagating from the SLM. The color-map in (a) shows the phase difference between the $E_x$ and $E_y$ orthogonal polarization components of the vector beam. The polarization states in (b) are plotted on top of the simulated engineered beam intensity. The black and blue colors in (b) represent the opposite handedness of the polarization ellipses. The polarization singularity structures such as lemon, star and L-line are also marked. Adapted with permission from [9] © The Optical Society.

on an array sensor placed in the back focal plane of the Fourier transforming lens. Let us denote the transmission function of the phase screen in the lens aperture by $t(x, y)$, then the total intensity of the engineered beam as observed on the CCD sensor in the back focal plane of the lens may described (apart from constants) as:

$$I_{\text{Total}}(u, v) = I_1 + I_2 = \left| \mathcal{F}\{t(x, y)E_x(x, y)\} \right|^2 + \left| \mathcal{F}\{t(x, y)E_y(x, y)\} \right|^2, \quad (9.3)$$

where $E_x$ and $E_y$ denote the $x$ and $y$ components of the field in the random phase screen plane. It is worth noting that the electric fields corresponding to the $x$ and $y$ polarizations in the camera plane essentially have a two-dimensional Hilbert transform relationship due to the OAM diversity in $E_x$ and $E_y$. The coordinates in the camera plane are denoted by $(u, v)$. The operation $\mathcal{F}$ denotes the two-dimensional Fourier transform operation corresponding to the Fraunhofer diffraction. In writing the total intensity as a sum of the two polarizations, we have neglected the cross-talk between the two orthogonal polarization components as it is usually negligible in the case of thin random phase screens [10] and becomes important only when dealing with thick scattering media. The intensity patterns $I_1$ and $I_2$ for individual polarizations may be observed on the sensor by introducing a polarizer in the beam path. The experimental intensity profiles corresponding to the simulated intensity patterns shown in figure 9.2 are shown in figure 9.5. In this experiment, crumpled transparent polythene sheets were used as random phase screens [9]. The smooth total intensity patterns for various random phase screens shown in figure 9.3 suggest that examining the diversity in $I_1$ and $I_2$ may be an interesting method for robust beam design as we will explore in the next section.

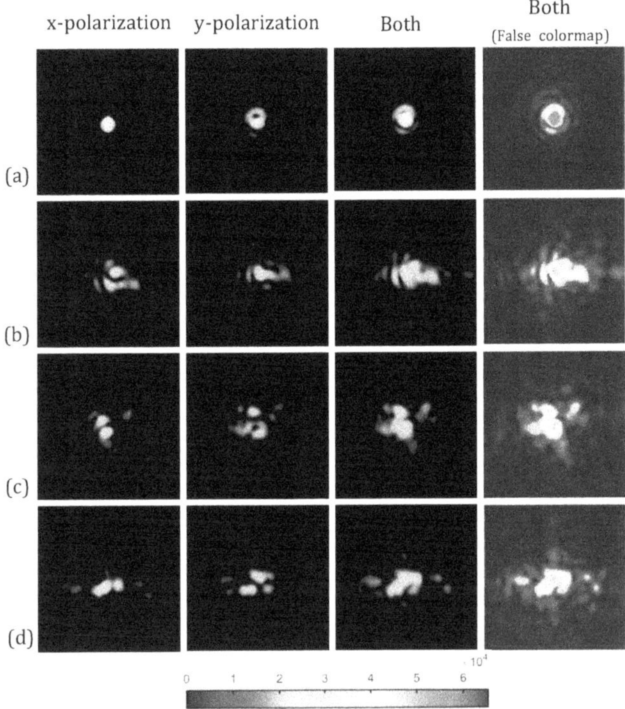


**Figure 9.5.** The experimentally recorded intensities for vector beam along with its constituent orthogonal polarizations are shown for three different realizations of the random phase screen in (b)–(d). Here, different regions of a crumpled polythene sheet have been used as the random phase screens. The first row (a) shows the recorded intensities when the beam passes through open aperture. The last column shows the total intensity of the vector beam in false colors. Adapted with permission from [9] © The Optical Society.

## 9.2 Beam quality assessment using instantaneous SNR

We have discussed the concept of speckle diversity for robust beam design for propagation through random phase perturbations. In this section, we introduce a beam quality parameter—the instantaneous SNR—which provides quantitative assessment of the improvement in beam intensity profile. As we will discuss later, the on-axis scintillation index (SI) has commonly been used as a measure for beam quality typically for Gaussian beams. The SI, however, does not reflect the nature of improvement in the transverse beam profile. We define instantaneous beam SNR as the mean intensity $\langle I \rangle$ divided by the standard deviation of intensity $\sigma_I$ values within a given detector area:

$$\mathrm{SNR}(r_d) = \frac{\langle I \rangle}{\sigma_I}.$$

(9.4)

The parameter $r_d$ denotes the radius of the detector area that has been used for computing the SNR. The SNR is calculated for individual realizations of the random phase screens which may then be averaged over the total number of

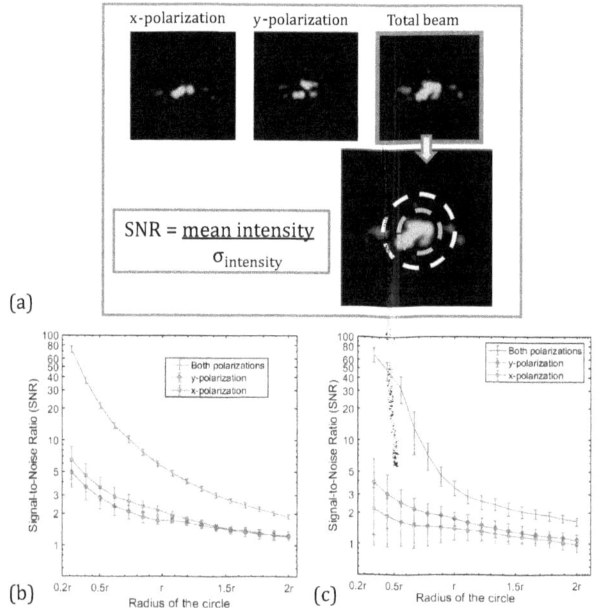


**Figure 9.6.** The methodology for the calculation of SNR for the engineered beam is explained in (a). The point where the intensity of the vector beam is maximum is located and around this point, a circular area of variable radius is considered. The SNR is calculated for the intensity pixels which fall inside this circular region. The variation of SNR with respect to the radius of the circle is shown in (b) for the simulated intensity images in figure 9.2, and (c) for the experimentally obtained intensity images respectively. Adapted with permission from [9] © The Optical Society.

realizations. For calculation of SNR, we first find the location of the peak intensity of the engineered vector beam. Then, a circular area centered at this peak location is selected from the beam and the SNR is calculated for the enclosed intensity values. These circular regions have been centered on the peak position of the engineered beam as this is naturally the best position of the detector in any real practical system. Typically fast kHz rate tip–tilt corrections in the beam may be made to bring the peak intensity to the detector center. Figure 9.6 shows the SNR curves for individual polarizations as well as total beam intensity for simulated (figure 9.2) as well as experimental (figure 9.5) intensity profiles. The parameter $r$ on the $x$-axis of the SNR plots represents the radius of the free space diffraction-limited spot for the lens-camera Fourier transform arrangement in figure 9.3. The error bars in these curves represent standard deviations in the SNR values corresponding to the 50 realizations of the random phase screens used in simulations and experiments respectively. It is observed that the SNR values for the engineered beam are consistently greater than those for the individual polarization components in both simulation and experiments. At a detector radius of half the diffraction-limited spot, the beam SNR is observed to be higher by an order of magnitude or more which may offer a significant advantage in terms of maintaining a smooth beam profile over the detector.

## 9.3 Speckle diversity

The recent literature has hinted at the advantage of using inhomogeneously polarized beams for propagation through turbulence [9, 11–17]. As discussed in chapter 6, inhomogeneously polarized light can be considered to be a vector superposition of distinct amplitude-phase wave functions in the two polarizations. The two orthogonally polarized states of an inhomogeneously polarized beam do not interfere and as a result, the overall intensity profile is equal to the sum of its two polarization components. If these two polarization states evolve into sufficiently diverse speckles on propagation through a random medium, then their sum intensity will have a smoother profile compared to the individual polarization states. Keeping this diversity idea in mind, we explore the propagation of vector beams in various states of polarization through a random phase screen and observe their far-field diffraction patterns. We will specifically study the C-point and V-point polarization singular beams as they form an important class of inhomogeneously polarized beams.

Consider a vector beam propagating in the $+z$ direction where its electric field may be described in the paraxial form as,

$$E(x, y, z) = [E_1(x, y)\hat{e}_1 + E_2(x, y)\hat{e}_2]\exp(ikz). \tag{9.5}$$

The orthogonal unit vectors $\hat{e}_1$ and $\hat{e}_2$ may denote Cartesian vectors $(\hat{x} - \hat{y})$ denoting linear polarization or the right circularly or left circularly polarized (RCP–LCP) states $(\hat{e}_R - \hat{e}_L)$. Here, a time dependence of the form $\exp(-i\omega t)$ and circular polarization basis $(\hat{e}_1 = \hat{e}_R, \hat{e}_2 = \hat{e}_L)$ has been assumed. The polarization distribution of the vector beam is decided by the functional form of the amplitude and phase of the two orthogonal polarization components $E_1(x, y)$ and $E_2(x, y)$. When these two components have different amplitude and phase structure, the resulting vector beam has spatially inhomogeneous polarization. Our aim is to understand if we can design $E_1(x, y)$ and $E_2(x, y)$ such that the total intensity pattern $I_{Total}(u, v) = I_1(u, v) + I_2(u, v)$ as in equation 9.3 can be made resistant to random phase perturbations. In this context, the diversity in the speckle pattern intensities $I_1(u, v)$ and $I_2(u, v)$ is an important factor to be taken into consideration. Clearly if $I_1(u, v)$ and $I_2(u, v)$ are highly correlated, their addition will not lead to much gain in terms of reducing the intensity fluctuations. The traditional literature on speckle has considered reduction in speckle contrast on addition of uncorrelated speckles [18]. However, if the two speckle patterns $I_1(u, v)$ and $I_2(u, v)$ are negatively correlated, the resulting total intensity pattern would have lesser contrast and more uniformity. This is so because in negatively correlated speckle patterns, the location of intensity maxima for one polarization will coincide with the location of intensity minima for the other. Therefore, the addition of such complementary speckles would reduce the fluctuations in the overall intensity. Now, we study four different types of vector beams with different amplitude and phase structures for $E_1(x, y)$ and $E_2(x, y)$. These beams have been chosen such that the speckles corresponding to the $E_1(x, y)$ and

$E_2(x, y)$ components span a range of correlation coefficients from $+1$ to $-1$. This is achieved by constructing vector beams with increasingly different $E_1(x, y)$ and $E_2(x, y)$ profiles in terms of their amplitude, phase and OAM. These vector beams are then propagated through a random phase screen and the correlation coefficient (see equation (9.1)) between the intensities $I_1(u, v)$ and $I_2(u, v)$ is calculated over the diffraction-limited spot size. The simulated far-field intensities of the vector beams and their orthogonal polarization components ($\hat{e}_R - \hat{e}_L$) are illustrated in figure 9.7 for a given realization of the random phase screen. The first column in the figure provides the polarization distribution of the vector beam. The mean correlation coefficient is mentioned in the last column which has been calculated over a set of 50 independent realizations of the random phase screen. We next describe these four cases one by one.

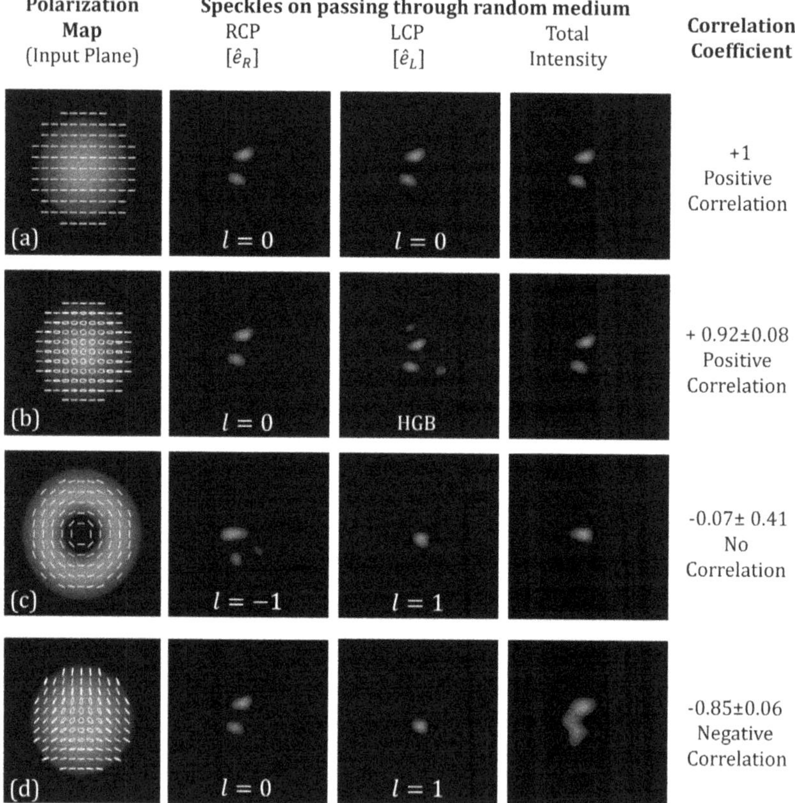

**Figure 9.7.** Simulation results of beams with different polarization structure and their corresponding far-field intensity profile on passing through a random phase screen. The intensities $I_1$ and $I_2$ for the RCP and LCP components of the beams are also shown. The correlation coefficient $\rho_{I_1,I_2}$ has been averaged over 50 random phase screen realizations. Note that here HGB stands for hollow Gaussian beams. Reprinted with permission from [12] © 2018 by the American Physical Society.

1. **Scalar beam**: The first case shown in figure 9.7(a) corresponds to a homogeneously polarized beam with equal energy in the $\hat{e}_R$ and $\hat{e}_L$ components. Both $E_1(x, y)$ and $E_2(x, y)$ components are taken in the $l = 0$ OAM state. In this case, the speckles produced by the two orthogonal polarizations on passing through a given random phase screen are identical with perfect correlation of $\rho_{I_1,I_2} = +1$. Therefore, when the two intensities $I_1$ and $I_2$ are added, there is no advantage in terms of beam quality.

2. **Amplitude diversity**: Next, we consider the case of an inhomogeneously polarized beam obtained by using different amplitude profiles for $E_1(x, y)$ and $E_2(x, y)$ components as shown in figure 9.7(b). The $\hat{e}_R$ polarization contains a $l = 0$ OAM mode while the $\hat{e}_L$ polarization contains a dark hollow Gaussian beam (HGB) [4, 19]. The HGB has a donut intensity profile similar to the $l = 1$ OAM mode but it does not carry any OAM. The HGB field at $z = 0$ plane can be expressed as:

$$E_{HGB} = A_o \exp(-r^2/w_o^2)(r^2/w_o^2)\exp(ikz), \tag{9.6}$$

where $r = \sqrt{x^2 + y^2}$ is the radial coordinate in the transverse plane, $w_o$ is the beam waist and $A_o$ is the normalization constant of the $l = 0$ state. Therefore, the vector beam can be written as,

$$\boldsymbol{E}(x, y) = A_o \exp(-r^2/w_o^2) [\hat{e}_R + (r^2/w_o^2)\hat{e}_L]\exp(ikz). \tag{9.7}$$

For this beam, we find that the intensities $I_1$ and $I_2$ are positively correlated with a correlation coefficient value of $\rho_{I_1,I_2} = 0.92 \pm 0.08$. Therefore, the vector beam intensity profile would show similar fluctuations as observed for the individually polarized components. This means that any inhomogeneously polarized beam formed as a result of only amplitude diversity in the two polarizations components is not necessarily useful in maintaining a robust intensity through turbulence.

3. **Phase diversity**: For the third case, we add phase diversity in the two polarization states in the form of different OAM states $l = -1$ and $l - 1$, given by $\exp(-i\theta)$ and $\exp(i\theta)$ respectively. The two polarization states now contain phase singularity of the same strength but opposite handedness. Embedding the $l = -1$ and $l = 1$ states in the $\hat{e}_R$ and $\hat{e}_L$ polarization states lead to the V-point polarization singularity in the form of the azimuthal polarization. The intensity profiles for this beam and its constituent orthogonal polarizations is shown in figure 9.7(c). It is observed that the correlation coefficient between the intensity profiles $I_1$ and $I_2$ is $\rho_{I_1,I_2} = -0.07 \pm 0.41$ which indicates uncorrelated speckle patterns on an average. One can see from the figure 9.7(c) that the speckle contrast in this case reduces for the vector beam as compared to the individual $\hat{e}_R - \hat{e}_L$ states.

4. **OAM diversity**: For the last case, we have used $l = 0$ and $l = 1$ OAM states in the $\hat{e}_R$ and $\hat{e}_L$ polarizations. This choice of OAM states in the orthogonal polarization produces a C-point polarization singular structure of star type in this case. This beam is essentially the same as shown in figure 9.3 except that

instead of the $(\hat{x} - \hat{y})$ polarization basis given by polarizations the current beam has a circular polarization basis. Figure 9.7(d) shows the intensity profiles of the beam and its components. We observe that the intensity profile of the vector beam has significantly improved compared to the $l = 0$ and $l = 1$ OAM states. This observation is also supported by the correlation coefficient value $\rho_{I_1,I_2} = -0.85 \pm 0.06$ which shows high negative correlation between the $I_1$ and $I_2$ profiles. The high negative correlation can also be seen in the $I_1$ and $I_2$ intensity patterns where the maxima of one polarization complements the minima of the other polarization. This complementarity property is unique to C-point structures. This complementary evolution of the two polarization structures can thus keep the total intensity of the beam robust even in a time-varying random medium.

The polarization singularity structures described above may be generated in the laboratory using the methods described in chapter 6 by vector superposition of scalar waveforms in orthogonal polarizations. The above analysis shows that not all inhomogeneously polarized beams are equally robust against phase perturbations. Only specific designs of the vector beams whose orthogonal polarizations produce either uncorrelated or negatively correlated speckle patterns offer an improvement in the total intensity profile of the vector beam. Therefore, an intelligent choice of $E_1(x, y)$ and $E_2(x, y)$ should be made in the vector beam design. In this regard, the speckle diversity in the intensity profiles of the orthogonal polarization is an important feature which can be used as a general guideline for designing robust beams.

The experimentally recorded intensities [12] for the V-point and C-point singular beams after they have passed through the random phase screen are shown in figure 9.8. The panels (a) and (b) in the figure corresponds to V-point singular beams, namely the azimuthal and radial polarizations while the panels (c) and (d) show the lemon and star type C-point polarization singular beams respectively. The decomposition of these beams in circular basis is also illustrated diagrammatically. The figure shows the intensity $I_{\text{Total}}$ of the vector beams and the intensities $I_1$ and $I_2$ of their two constituent polarization components. In each of the panels (a)–(d), row I shows the intensities for the beams as they pass through an open aperture without any random phase screen while row II shows the intensity records when a random phase screen was present in the aperture plane. It can be easily seen through visual inspection that the beam quality of the vector beams is superior when it contained a C-point polarization singular structure as compared to the cases when it had a V-point polarization singular structure. The correlation coefficient calculated between the experimentally recorded $I_1$ and $I_2$ intensities for the beams is shown in table 9.1. The standard deviation in the correlation coefficient is calculated over 50 experimental realizations of the random phase screen. The quality of the V-point and C-point singular beams after propagation through the random medium can be quantified using the beam SNR. The SNR curves for experimental and simulation data [12] have been shown in figure 9.9. Panels (a) and (b) show the SNR curves for

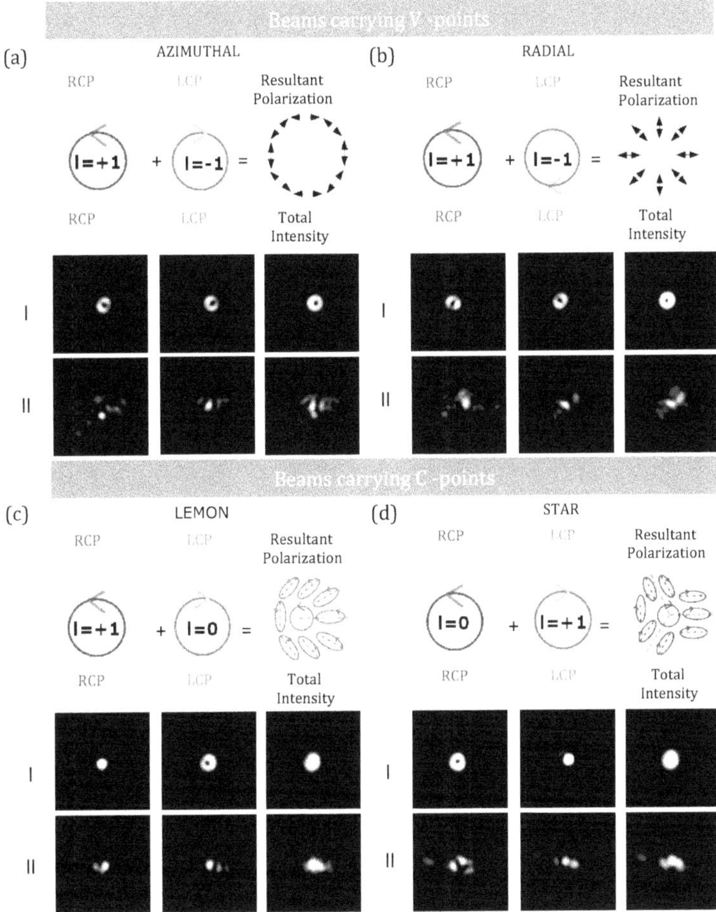


**Figure 9.8.** The experimentally obtained intensities for beams carrying V-point and C-point polarization singularity are shown in (a), (b) and (c), (d) respectively. Here, (a) and (b) show the azimuthal and radial V-point singular beams while (c) and (d) illustrate the cases of lemon and star type C-point polarization singular beams. The decomposition of these beams in RCP–LCP basis is also provided. Row I shows the Fraunhoffer diffraction intensities for the no-phase screen case while row II shows the intensities when the beam has passed through a random phase screen. Reprinted with permission from [12] © 2018 by the American Physical Society.

**Table 9.1.** Correlation coefficient for the $I_1$ and $I_2$ intensity patterns on CCD sensor for five different types of polarization singular beams.

| Type of singularity | Singularity structure | Correlation coefficient |
| --- | --- | --- |
| C-point | Lemon | $-0.81 \pm 0.06$ |
| C-point | Star | $-0.76 \pm 0.07$ |
| C-point | Lemon–star dipole | $-0.82 \pm 0.06$ |
| V-point | Radial | $0.11 \pm 0.41$ |
| V-point | Azimuthal | $0.19 \pm 0.35$ |

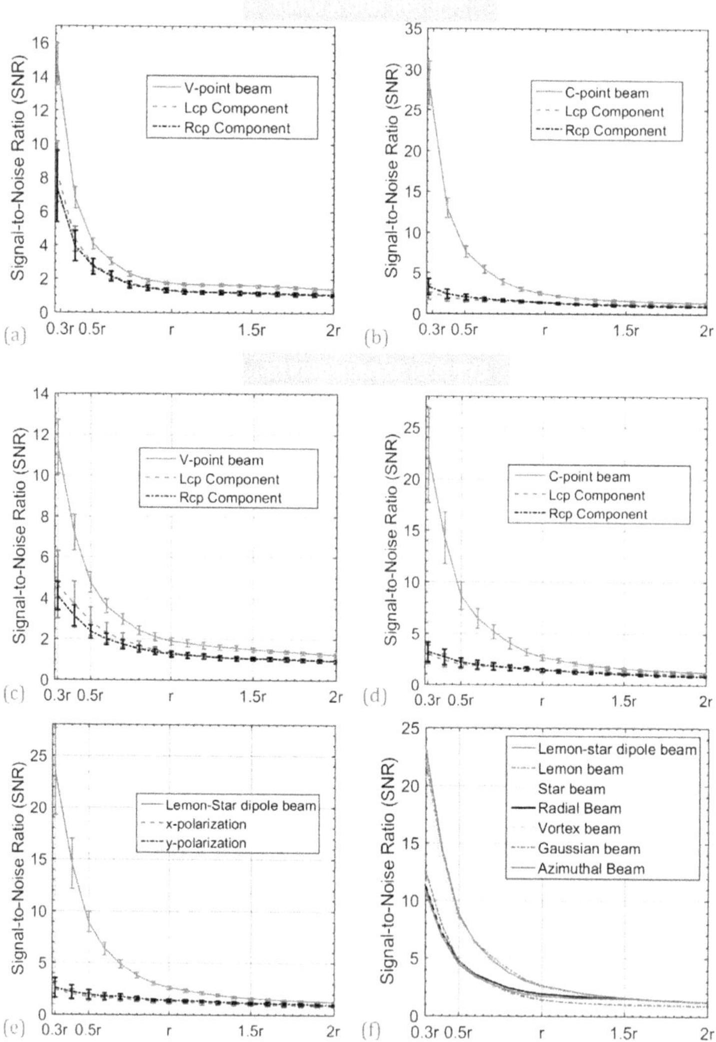

**Figure 9.9.** Simulated and experimentally observed SNR curves for the total beam and its RCP–LCP components are shown for V-point and C-point beams in panels (a) and (c) and panels (b) and (d) respectively. Panel (e) shows SNR curves for beam containing C-point dipole structure. Panel (f) shows a summary SNR comparison for C-point carrying beams, V-point carrying beams, C-point dipole beam and homogeneously polarized Gaussian and a charge +1 vortex beam. Reprinted with permission from [12] © 2018 by the American Physical Society.

the simulated intensities of the V-point and the C-point polarization singular beams. These plots show the variation of the SNR values as a function of the circular region used for calculation of SNR. The parameter $r$ on the $x$-axis of the plots corresponds to the radius of the free space diffraction-limited spot size. Once again, the error bars in these plots represent the standard deviation of the SNR values calculated over 50 realizations of the random phase screen. The SNR curves for the experimentally obtained intensities of the V-point and C-point polarization singular beams is shown in panels (c) and (d) respectively. Panel (e) shows the SNR values for a C-point polarization singular beam. The polarization distribution of this beam is in the form of a lemon–star C-point dipole. The last panel (f) contains a summary of the experimentally obtained SNR results for various V-point and C-point beams for convenience. The SNR values for the scalar $l = 0$ and $l = 1$ OAM states is also shown. One main difference in this plot is that the SNR calculation for the $l = 0$ and $l = 1$ OAM beams has been carried out at their individual intensity peak position. Note that there is a slight difference in the SNR values for the $\hat{e}_R$ and $\hat{e}_L$ polarized components in the V-point and C-point polarization singular beams in figures 9.9(a)–(d). The peak intensity in the V-point polarization singular beam typically occurs near the intensity peak of either one of its two constituent polarizations. This is understandable as the individual polarizations of the V-point singular beam containing $l = -1$ and $l = +1$ OAM states produce uncorrelated speckle patterns as observed from the table 9.1. Therefore, during the calculation of SNR values, the center of the circular region used for calculating SNR, is closer to the peaks of the $I_1$ or $I_2$ patterns, thus yielding a slightly different SNR value for the RCP–LCP components. However, for the C-point beam, the intensity patterns $I_1$ and $I_2$ are negatively correlated, therefore the peak intensity for the C-point polarization singular beam is typically not close to the peak intensities of either $I_1$ and $I_2$ patterns. Therefore, the center of the circular region used for calculating SNR is close to none of the peak values of $I_1$ and $I_2$ patterns, thus slightly decreasing the measured SNR values.

We make some important observations from the SNR plots. First of all, the SNR for the $I_{\text{Total}}$ is always more than that for the individual polarization intensities $I_1$ and $I_2$ for both V-point and C-point polarization singular beams. The C-point polarization singular beam shows approximately two-fold or more SNR gain over the V-point polarization singular beams when the SNR is calculated over a circle of size $0.5r$, $r$ being the radius of the diffraction-limited spot size. The better beam quality for C-point singular beams is seen irrespective of the polarization basis used as expected. The comparison of various scalar and vector beams in panel (f) highlights the importance of choosing suitable $E_1(x, y)$ and $E_2(x, y)$ fields for engineering robust vector beams. Finally, the complementary speckles of $l = 0$ and $l = 1$ OAM states is a convenient way of increasing speckle diversity in the $I_1$ and $I_2$ profiles which in turn makes sure that the overall beam intensity has reduced fluctuations on experiencing fluctuating random medium such as atmosphere.

## 9.4 Long range propagation of converging polarization singularities through atmospheric turbulence

In all the illustrations in this chapter so far, the random phase fluctuations were modeled using a single random phase screen. In practice, when propagation through long range turbulence (~few km) is of interest, a more appropriate model is to use the split-step method for computing field propagation as explained in chapter 8. In particular, a number of random phase screens with known power spectrum are generated and placed along the beam path interspersed with segments of free space propagation. In long range propagation, it is a common practice to introduce a concave curvature at the input side in order to focus the beam on the target. While the split-step method is well known, our main goal is to understand if our idea of complementary diffraction for the $l = 0, 1$ OAM states still holds when the long range propagation is modeled with this more rigorous approach. The material provided here is largely based on the work of the present authors [20].

A number of theoretical and simulation studies of long range propagation in atmospheric turbulence have been undertaken for Gaussian beams [21, 22]. They have been investigated in different turbulence strengths, beam parameters and focusing conditions and also as a potential ground station to satellite uplink [23–26]. Various other beam amplitude types like Bessel [27], Laguerre–Gaussian, Annular, cosh and cos Gaussian [2] have also been investigated. Other types of Gaussian beams like the dark hollow Gaussian [4] and flat-top Gaussian [3] have also been studied. The phase singularities of optical beams are known to be resistant against phase and amplitude perturbations [28]. The topological charge of vortex beams has been shown to be conserved on propagation through atmosphere [29] and has been useful as an information carrier in free space optical communication. Different variants of vortex beams like the flat-topped vortex beams and different turbulence strengths have also been studied [30–32]. Propagation of OAM beams have also been explored in the context of maintaining entanglement in quantum communication [33–36]. It has been observed that the scintillation for partially polarized and partially coherent beam is less than that of a fully coherent beam [5–8]. It has been theoretically shown by Schulz [37] that beams which have minimum scintillation are in general partially coherent. Many optimization schemes have been developed to incorporate these beams in free space optical communication [38]. The use of vector beams with inhomogeneous polarization has been proposed [11, 12, 39]. An inhomogeneously polarized beam can be expressed as a coherent superposition of two orthogonally polarized spatial modes. In this respect, vector vortex beams with radial and azimuthal polarization have been studied [13–15, 40].

Polarization singular beams have been described in detail in chapter 6. The lowest order C-point and V-point beams can be written in the form:

$$\vec{E}_{\mathrm{C}}(\boldsymbol{r}, z) = \frac{1}{\sqrt{2}}[re^{i\theta}\hat{e}_{\mathrm{R}} + \hat{e}_{\mathrm{L}}]\psi(\boldsymbol{r}, z), \tag{9.8}$$

$$\vec{E}_{\mathrm{V}}(\boldsymbol{r}, z) = \frac{1}{\sqrt{2}}[re^{i\theta}\hat{e}_{\mathrm{R}} + re^{-i\theta}\hat{e}_{\mathrm{L}}]\psi(\boldsymbol{r}, z), \tag{9.9}$$

where $\psi(\boldsymbol{r}, z)$ is the host beam profile given by the Gaussian or the LG(0,0) mode. Therefore, the C-point beam is simply a linear combination of LG(0,1) and LG(0,0) modes in circular basis whereas the V-point beam is the combination of LG(0,1) and LG(0,−1). The host beam $\psi(\boldsymbol{r}, z)$ may therefore be expressed as

$$\psi(r; z = 0) = A \exp\left(-\frac{r^2}{W_o^2} - \frac{i\pi r^2}{\lambda F_o}\right), \tag{9.10}$$

where $A$ is the complex amplitude and $\lambda$ the wavelength. The beam waist of the Gaussian amplitude distribution is denoted by $W_o$ at the $z = 0$ plane while the radius of curvature of the wavefront is given by $F_o(>0)$. The beam waist at any distance $z > 0$ may be computed using Fresnel propagation and is given by [41]:

$$W(z) = W_o\sqrt{\left(1 - \frac{z}{F_o}\right)^2 + \left(\frac{z\lambda}{\pi W_o^2}\right)^2}. \tag{9.11}$$

By differentiating equation (9.11) with respect to $z$, the location $z_f$ at which the beam would nominally focus in free space can be obtained as:

$$z_f = F_o\left[1 + \left(\frac{2F_o}{kW_o^2}\right)^2\right]^{-1}. \tag{9.12}$$

It can be observed from equation (9.12) that $z_f < F_o$ which means that the effective focal length is shorter than the radius of the curvature of the beam [21, 42]. For a beam with $W_o$ of the order of tens of cm and $F_o$ value in km, this change in the focusing length is not negligible. For example, for a beam with $W_o = 7$cm, $\lambda = 1.55\,\mu$m and $F_o = 2000$ m, $z_f = 1922$ m. In order to focus the beam at a particular $z$ distance, equation (9.12) can be inverted and solved for $F_o$. This yields the expression for $F_o$ as,

$$F_o = \frac{k^2 W_o^4}{8z_f}\left[1 - \sqrt{1 - \frac{16z_f^2}{k^2 W_o^2}}\right]. \tag{9.13}$$

The result in equation (9.13) is true in free space, however, the beams behave differently in turbulence. An interesting observation that has been reported is that the focus spot moves slightly toward the transmitter [23]. In the current study, this focus shift has not been considered and all the beam parameters like SNR, SI, etc, are calculated at the final propagation plane situated at distance $z = 2$ km. The geometry of propagation is as shown in figure 8.2 except that now the input beam has additional concave curvature. In the following illustrations we use von Kárman spectrum for representing the refractive index fluctuations with inner scale $l_o = 0.01$ m and outer scales $L_o = 3$ m. The refractive index structure constant is taken as $C_n^2 = 10^{-14}$ m$^{-2/3}$ and $C_n^2 = 10^{-13}$ m$^{-2/3}$ for representing moderate and high level turbulence while the propagation distance $z$ is kept equal to 2 km. The various parameters used for simulation are listed in table 9.2.

### 9.4.1 Evolution of intensity and polarization structure of the beams

The total intensity for the polarization singular beam on propagation can be obtained by adding individual intensities of $\hat{e}_R$ and $\hat{e}_L$ polarizations. This is

Table 9.2. Simulation parameters for propagation of polarization singular beams through atmospheric turbulence.

| Parameter name | Symbol | Value |
| --- | --- | --- |
| Propagation distance | $z$ | 2000 m |
| Radius of curvature | $F_o$ | 2019 m |
| Inter-screen distance | $\delta z$ | 100 m |
| Number of screens | $M$ | 20 |
| Grid points | $N$ | 512 |
| Grid spacing | $\delta x$ | 0.001 m |
| Beam waist | $W_o$ | 0.1m |
| Wavelength | $\lambda$ | 1.55 $\mu$ m |
| Inner scale | $l_o$ | 0.01 m |
| Outer scale | $L_o$ | 3 m |

approximately true for media like atmosphere where the cross-talk between the two orthogonal polarization components may be neglected in low angle forward propagation. The total intensity profile of C-point and V-point polarization singular beams at different propagation distances has been shown in figures 9.10(a) and 9.11(a) for $C_n^2 = 10^{-14}$ m$^{-2/3}$. These figures show the intensity evolution of these two beams for a given realization of the atmospheric turbulence. It can be seen that at $z = 2$ km, the instantaneous intensity profile of the C-point beam is more uniform than that of the V-point beam. The polarization structure of these beams along with the corresponding Stokes phase is also illustrated in figures 9.10(b), (c) and 9.11(b), (c) respectively. The Stokes phase (as described in chapter 6) is characteristically different for V-point and C-point polarization singular structures and therefore, it offers a great visual aid for identification of the polarization singular structures in the beam. For example, as shown in figures 9.10(c) and 9.11(c), we observe that for a C-point beam, the Stokes phase has only one phase cut and the total phase variation is from 0 to $2\pi$ while the Stokes phase for the V-point beam has two phase cuts each going from 0 to $2\pi$. The C-point beam shown here has a lemon type polarization singular structure in the initial plane while the V-point beam has an azimuthal polarization structure.

A few observations can be made from the polarization and Stokes phase distributions of these beams on propagation through long range turbulence. The polarization distribution of the C-point beam shows a clockwise rotation as $z$ increases as seen in figure 9.10(b). This has been reported earlier for the lowest order C-point beam when propagated in free space [43] and in linear and non-linear media [44] where the polarization distribution is seen to undergo a rigid rotation of $\pi/2$ as it propagates from waist plane to the far-field zone. In figure 9.10(c), the Stokes phase also shows this clockwise rotation of the polarization pattern. The Stokes phase distribution of the V-point beam as seen in figure 9.11(c), shows that near $z = 1250$ m,

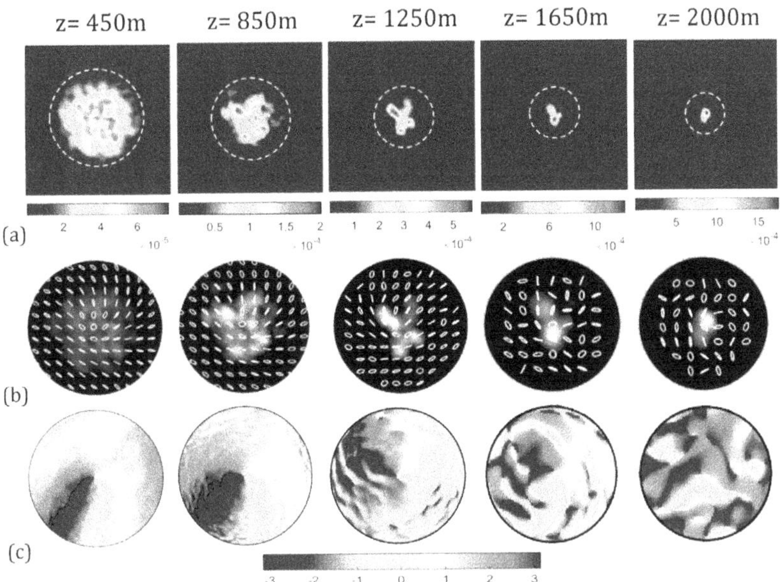



**Figure 9.10.** The figure shows the evolution of the C-point polarization singular beam on propagation through moderate turbulence levels denoted by $C_n^2 = 10^{-14}$ m$^{-2/3}$. Here, (a) shows the evolution of intensity profile of the beam for different $z$ values. Here each window size is 320 ×320 pixels. The polarization pattern and the corresponding Stokes phase of the beam for the region lying inside the dotted white circle is shown in (b) and (c) respectively. Note that the polarization pattern is shown superimposed on the beam's intensity profile. Adapted with permission from [20] © The Optical Society.

the two oppositely charged vortex components in the V-point probably splits into two separate vortices. At this point the polarization distribution is also seen to get scrambled. Due to the atmospheric turbulence, the two orthogonal polarization components of the V-point beam i.e. $l = 1$ and $l = -1$ OAM states evolve differently. This might result in a lateral shift in the position of the vortices of the $l = 1$ and $l = -1$ OAM states, thus resulting in two spatially separated cuts in the Stokes phase of the V-point beam. In general, we see that the polarization distribution gets scrambled for both V-point and C-point beams with increasing propagation distance. However, the Stokes phase structure of the C-point beam is preserved better compared to that of the V-point beam.

The intensity evolution and polarization distribution of C-point and V-point beams at different $z$ values is shown in figure 9.12 for the case when $C_n^2 = 10^{-13}$ m$^{-2/3}$. A few observations can be drawn from the illustrations. The beams experience more spreading due to higher phase fluctuations in this case. Our observation of the beam spot over multiple realizations shows that some portion of the C-point beam is still able to preserve a uniform intensity area within the spot while there is no corresponding uniform intensity area seen in the V-point beam for the same turbulence realizations. The intensity profiles of the beams at $z = 1650$m appear more focused than the intensity profiles at $z = 2000$ m. It is known that in turbulence, the beam focuses before its geometrical focus. However, this shifting of focus point

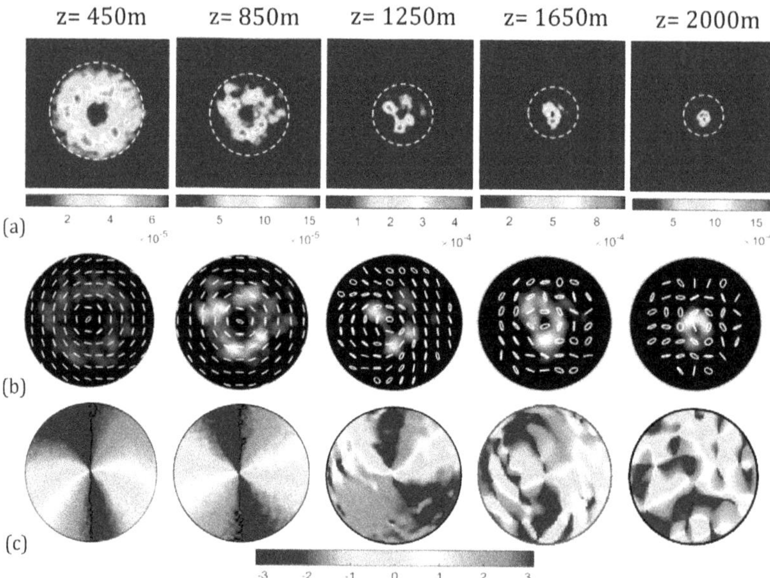


**Figure 9.11.** The figure shows the evolution of the V-point polarization singular beam on propagation through moderate turbulence levels denoted by $C_n^2 = 10^{-14}$ m$^{-2/3}$. Here, (a) shows the evolution of intensity profile of the beam for different $z$ values. Here each window size is 320 ×320 pixels. The polarization pattern and the corresponding Stokes phase of the beam for the region lying inside the dotted white circle is shown in (b) and (c) respectively. Note that the polarization pattern is shown superimposed on the beam's intensity profile. Adapted with permission from [20] © The Optical Society.

towards the transmitter seems to increase with increasing $C_n^2$ value. Note for comparison, the beam sizes in figures 9.10 and 9.11 where we see that for $C_n^2 = 10^{-14}$ m$^{-2/3}$, the beams focus point is near the geometrical focus situated at 2000 m whereas this is not true when $C_n^2$ is taken to be $10^{-13}$ m$^{-2/3}$. Therefore, in higher turbulence, the radius of curvature of the beam might need to be adjusted. Figure 9.12(c) and (d) show the polarization distribution and the corresponding Stokes phases for the beams at $z = 450$ m and $z = 850$ m respectively. It is clearly visible that the polarization structure of the beam is severely distorted by $z = 850$ m.

### 9.4.2 Quantitative assessment of beam quality

In order to quantitatively measure the beam quality, we next calculate the SNR and on-axis scintillation index (SI). Earlier studies on focused non-singular beams have shown that they have lesser scintillation near the focus point, however, these beams are more susceptible to beam wander [25, 26, 45–48]. Therefore, in order to have any actual reduction in scintillation on the receiver, it is necessary that these beams should be actively corrected for beam wander. Therefore, we report SNR and on-axis SI for both cases of tracked and untracked beams. The beam position is usually centered onto the detector by using a tip–tilt correction to counteract the effects of beam wander. Fast scanning mirrors with kHz tip–tilt correction rate may be required for correcting the beam wander.

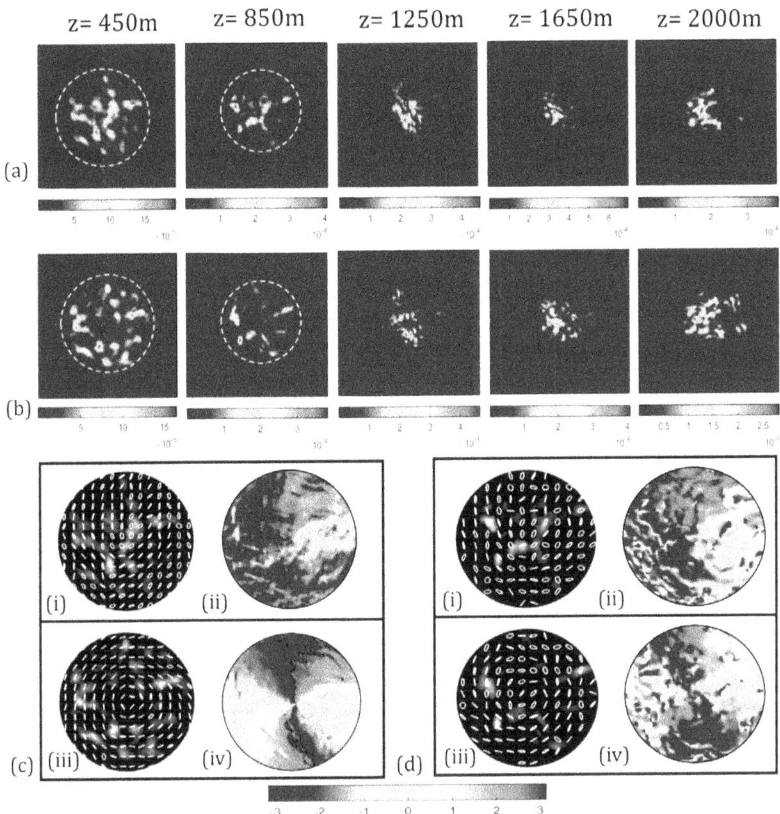

**Figure 9.12.** The figure shows the evolution of the intensity and polarization structure of the polarization singular beams on propagation through high turbulence levels denoted by $C_n^2 = 10^{-13}$ m$^{-2/3}$. Here (a) and (b) show the evolution of intensity profile of the C-point and V-point polarization singular beam for different $z$ values. Here each window size is 320 ×320 pixels. The polarization pattern and the corresponding Stokes phase of the beams at $z = 450$ m and $z = 850$ m for the region marked by dotted white circle is shown in (c) and (d) respectively. In both (c) and (d), (i) and (ii) correspond to the C-point beam while (iii) and (iv) are for V-point beam. Note that the polarization pattern is shown superimposed on the beam intensity profile.

We use an on-axis circular detector for calculation of SNR for the untracked beams. In the case of tracked C-point and V-point beams, the circular detector is always centered on their instantaneous peak intensity positions. For the same positions of the detector, SNR values for the intensity profiles of the individual orthogonal polarization states are also calculated. We have varied the radius of the detector from 25% to 150% of the free space diffraction-limited spot-radius of the Gaussian beam as given by equation (9.11). The variation of SNR of the C-point and V-point beams with increasing detector radius is shown in figure 9.13(a), (b) for the untracked beams and in figure 9.13(c), (d) for the tracked beams when $C_n^2 = 10^{-14}$ m$^{-2/3}$. The parameter 'r' here signifies the free space diffraction-limited spot size of the Gaussian beam. The SNR values of the two constituting polarizations are also plotted. The error bars in the figures represent the standard deviation of the SNR values calculated over 1000 independent realizations of atmosphere. Each of these 1000 realizations used 20 random phase screens generated using the FFT method with sub-harmonic correction. The error bars are shown on the composite vector beam but not on individual polarizations in order to avoid overcrowding the plot. Figure 9.13 shows that the SNR values for the two polarization singular beams are consistently greater compared to their orthogonal polarization components. At $z = 2$ km, there is not much difference in the SNR values of LG(0,0) and LG(0,1) beams individually, however when added incoherently, a significant improvement in SNR of C-point beam is seen. Similarly, the SNR values of LG(0,1) and LG(0,−1) are almost identical. Adding them incoherently improves the SNR of V-point beam but not as much as in the case of the C-point beam. The SNR values of both C-point and V-point beams shows further improvement for the tracked case as compared to the untracked case since now the corresponding beam profile on the detector is better. Figure 9.14 shows the SNR curves for the tracked polarization singular beams (averaged over 1000 realizations) when $C_n^2 = 10^{-13}$ m$^{-2/3}$. It was observed that even in high turbulence, the C-point beam still has a spot which is uniform although the overall beam has experienced more spread. The V-point beam does not show such a feature for the same realizations of random phase screens. Therefore, the C-point beam is still able to maintain a higher SNR value as compared to the V-point beam or the scalar $l = 0$ and $l = 1$ OAM states. The correlation coefficient is calculated for the tracked polarization singular beams over variable sized areas (different radius) at the receiver plane ($z = 2$ km) and is tabulated in table 9.3. The correlation coefficient values for $C_n^2 = 10^{-13}$ m$^{-2/3}$ follow a similar trend as those for $C_n^2 = 10^{-14}$ m$^{-2/3}$. However, as for higher turbulence, beams breaks up into multiple speckles and experience a larger spread, therefore the variance in the measured values of $\rho_{l_1,l_2}$ is larger for $C_n^2 = 10^{-13}$ m$^{-2/3}$. In general, it is observed that the intensity patterns corresponding to the orthogonal polarization states of the C-point beam have significant negative correlation (correlation coefficient $<-0.5$) up to a sizable receiver radius for both moderate and high turbulence levels. On the other hand the orthogonal polarization states of the V-point beam are nearly uncorrelated for the same receiver sizes. The observed results obtained with multiple phase screen model for turbulence are in agreement with results when a single phase screen was used to model turbulence. The negative intensity correlation in the orthogonal polarization states corresponding to the $l = 0$ and $l = 1$ states is a feature that is unique to the C-point polarization singularity which is seen to be important for robust beam design.

Figure 9.15 shows the variation of the on-axis SI as a function of propagation distance for both untracked and tracked polarization singular beams for $C_n^2 = 10^{-14}$ m$^{-2/3}$. The beams have been corrected for beam wander by tracking the centroid of the intensity profile. It is important to note that the tilt caused in the two orthogonal polarization components of the vector beams due to the sub-harmonic terms is of a similar nature. Further it is noticed that the typical beam wander for the simulation

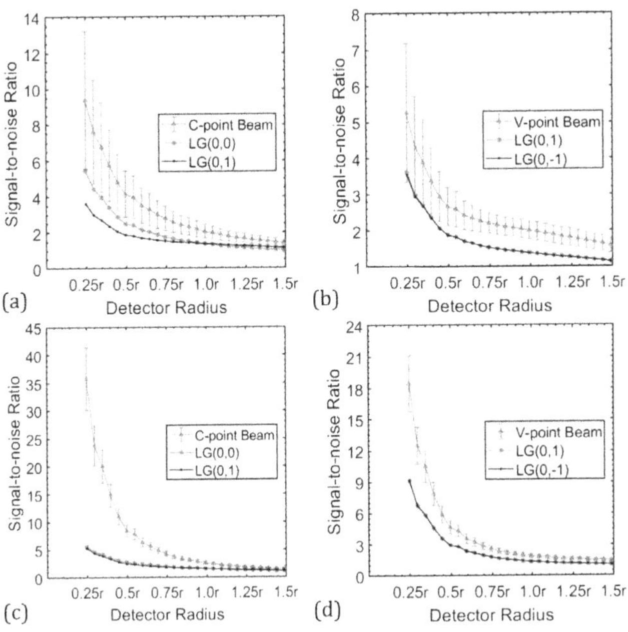

**Figure 9.13.** Variation of SNR with detector size for (a), (b) untracked C-point and V-point beams, and (c), (d) tracked C-point and V-point beams respectively for $C_n^2 = 10^{-14}$ m$^{-2/3}$. Adapted with permission from [20] © The Optical Society.

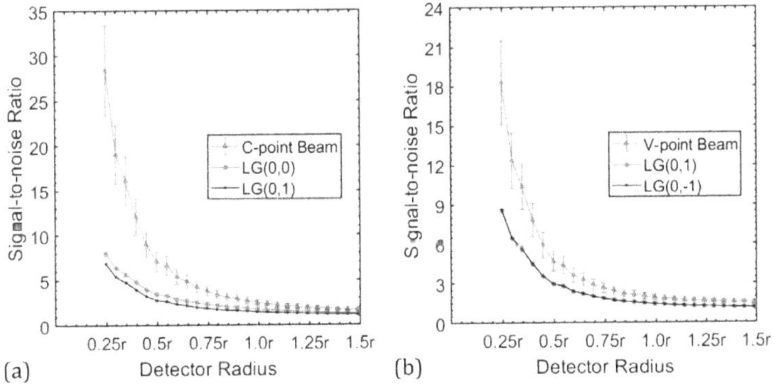

**Figure 9.14.** Variation of SNR with detector size for tracked (a) C-point and (b) V-point beams for $C_n^2 = 10^{-13}$ m$^{-2/3}$. Adapted with permission from [20] © The Optical Society.

parameters was seen to be up to 30% of the free space diffraction-limited spot size on average.

The traditional measure for assessing beam quality after propagation through turbulence is the scintillation index (SI). The on-axis SI is defined as:

$$\mathrm{SI}(0,\,0,\,z) = \frac{<I(0,\,0,\,z)>}{\sqrt{<I(0,\,0,\,z)^2> - <I(0,\,0,\,z)>^2}}. \tag{9.14}$$

**Table 9.3.** Correlation coefficient between the intensity patterns associated with orthogonal polarization components of C-point and V-point beams at $z = 2$ km for $C_n{}^2 = 10^{-14}$ m$^{-2/3}$ and $C_n{}^2 = 10^{-13}$ m$^{-2/3}$.

| Type of beam | Radius | Value of correlation coefficient $\rho_{I_1,I_2}$ | |
| --- | --- | --- | --- |
| | | $C_n{}^2 = 10^{-14}$ m$^{-2/3}$ | $C_n{}^2 = 10^{-13}$ m$^{-2/3}$ |
| C-point | 0.25$r$ | $-0.951\,4 \pm 0.046\,4$ | $-0.807\,3 \pm 0.201\,2$ |
| | 0.50$r$ | $-0.876\,7 \pm 0.076\,8$ | $-0.610\,9 \pm 0.290\,5$ |
| | 0.75$r$ | $-0.730\,5 \pm 0.120\,8$ | $-0.449\,4 \pm 0.312\,6$ |
| | 1.0$r$ | $-0.537\,1 \pm 0.159\,1$ | $-0.345\,1 \pm 0.316\,5$ |
| V-point | 0.25$r$ | $-0.505\,0 \pm 0.413\,4$ | $-0.463\,1 \pm 0.407\,8$ |
| | 0.50$r$ | $-0.144\,8 \pm 0.440\,2$ | $-0.251\,6 \pm 0.449\,6$ |
| | 0.75$r$ | $-0.134\,1 \pm 0.433\,8$ | $-0.175\,6 \pm 0.449\,0$ |
| | 1.0$r$ | $-0.195\,1 \pm 0.379\,9$ | $-0.177\,6 \pm 0.390\,1$ |

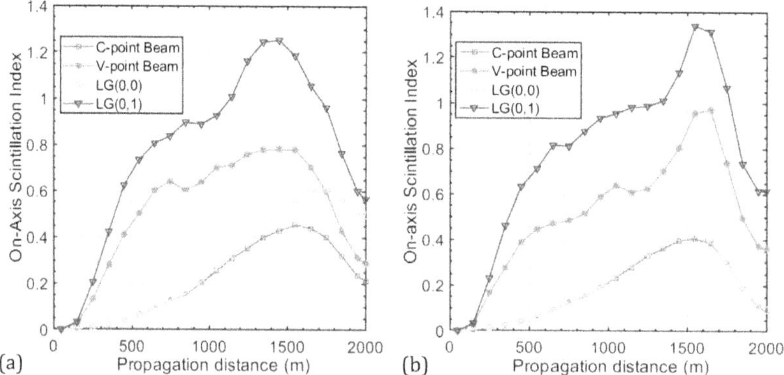


**Figure 9.15.** Variation of on-axis scintillation index with propagation distance for (a) untracked and (b) tracked beams for $C_n{}^2 = 10^{-14}$ m$^{-2/3}$. The markers represent the location of the random phase screens. Adapted with permission from [20] © The Optical Society.

Here the ensemble average is taken over a number of realizations of turbulence. The on-axis SI values have been calculated over a $3 \times 3$ pixel square aperture which is placed on-axis at $(x, y) = (0, 0)$ in the detector plane. The size of this square aperture is below $\sqrt{\lambda L/2\pi}$, which is the required condition for avoiding aperture averaging effects in SI calculation. The plots show the SI values averaged over 1000 turbulence screen realizations. The plots in figure 9.15 show that for both untracked as well as tracked cases, the on-axis SI values of the C-point beam are consistently lower than that for the V-point beam. For the tracked case, the on-axis SI for the C-point and LG(0,0) beam appears to be nearly identical over the propagation distances considered here. For the higher atmospheric turbulence case which is represented by $C_n^2 = 10^{-13}$ m$^{-2/3}$, the on-axis SI plots for the tracked beams is shown in figure 9.16. It can be seen that the C-point and V-point beams have comparable SI

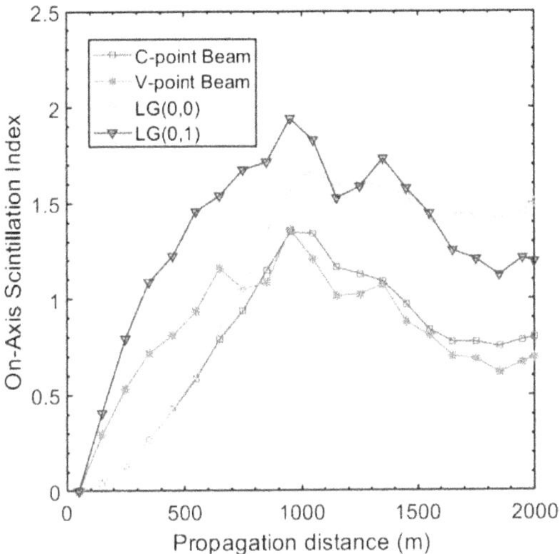



**Figure 9.16.** Variation of on-axis scintillation index with propagation distance for tracked beams for $C_n^2 = 10^{-13}$ m$^{-2/3}$. The markers represent the location of the random phase screens.

values in the strong turbulence case. The instantaneous SNR curves in figures 9.13 and 9.14, however, show that the SNR at $z = 2$ km for the C-point beam is significantly higher compared to the LG(0,0) beam. In on-axis SI calculations, one looks at the on-axis or central intensity of the beam, therefore for beams which naturally contain a null intensity at their core, like vortex beams, on-axis SI values are very high at low $z$ values. These beams on further propagation through turbulence develop a speckled appearance and experience larger wandering and beam spreading. As the long-averaged beam spot is obtained by summing over all the individual realizations, this process fills up the null at the center for untracked beams resulting in a decrease in SI values. Therefore, on-axis SI might be a biased approach for characterizing fluctuations in vortex beams and on-axis SI alone might not give correct insight into the beam behavior in turbulence. In this respect, measuring instantaneous SNR can be a valuable additional parameter to assess the quality of the spatial beam profile.

We have presented a robust beam engineering principle based on the complementary diffraction effect between diffraction patterns associated with $l = 0, 1$ OAM states. The robust beam design idea as presented here may have important implication for adaptive optics as well as free space communication systems. These topics need to be studied in detail in future.

# References


[1] Baykal Y, Eyyuboglu H T and Cai Y 2009 *Effect of Beam Types on the Scintillations: A Review* vol 7200 (Bellingham, WA: SPIE Optical Engineering Press) pp 7–21

[2] Eyyuboglu H T 2011 Annular, cosh and cos Gaussian beams in strong turbulence *Appl. Phys.* B **103** 763–9

[3] Cai Y 2006 Propagation of various flat-topped beams in a turbulent atmosphere *J. Opt. A: Pure Appl. Opt.* **8** 537–45

[4] Cai Y and He S 2006 Propagation of various dark hollow beams in a turbulent atmosphere *Opt. Express* **14** 1353–67

[5] Korotkova O 2008 Scintillation index of a stochastic electromagnetic beam propagating in random media *Opt. Commun.* **281** 2342–8

[6] Gbur G and Wolf E 2002 Spreading of partially coherent beams in random media *J. Opt. Soc. Am.* A **19** 1592–8

[7] Borah D K and Voelz D G 2010 Spatially partially coherent beam parameter optimization for free space optical communications *Opt. Express* **18** 20746–58

[8] Ricklin J C and Davidson F M 2002 Atmospheric turbulence effects on a partially coherent gaussian beam: implications for free-space laser communication *J. Opt. Soc. Am.* A **19** 1794–802

[9] Lochab P, Senthilkumaran P and Khare K 2017 Robust laser beam engineering using polarization and angular momentum diversity *Opt. Express* **25** 17524–9

[10] Wheelon A D 2001 *Electromagnetic Scintillation* vol 2 (Cambridge: Cambridge University Press)

[11] Wei C, Wu D, Liang C, Wang F and Cai Y 2015 Experimental verification of significant reduction of turbulence-induced scintillation in a full Poincaré beam *Opt. Express* **23** 24331–41

[12] Lochab P, Senthilkumaran P and Khare K 2018 Designer vector beams maintaining a robust intensity profile on propagation through turbulence *Phys. Rev.* A **98** 023831

[13] Cheng W, Haus J W and Zhan Q 2009 Propagation of vector vortex beams through a turbulent atmosphere *Opt. Express* **17** 17829–36

[14] Chen Z, Cui S, Zhang L, Sun C, Xiong M and Pu J 2014 Measuring the intensity fluctuation of partially coherent radially polarized beams in atmospheric turbulence *Opt. Express* **22** 18278–83

[15] Cox M A, Rosales-Guzmán C, Lavery M P J, Versfeld D J and Forbes A 2016 On the resilience of scalar and vector vortex modes in turbulence *Opt. Express* **24** 18105–13

[16] Gu Y, Korotkova O and Gbur G 2009 Scintillation of nonuniformly polarized beams in atmospheric turbulence *Opt. Lett.* **34** 2261–3

[17] Gu Y and Gbur G 2012 Reduction of turbulence-induced scintillation by nonuniformly polarized beam arrays *Opt. Lett.* **37** 1553–5

[18] Goodman J W 1975 *Statistical Properties of Laser Speckle Patterns* ed Laser Speckle (Berlin: Springer)

[19] Song Y, Milam D and Hill W T 1999 Long, narrow all-light atom guide *Opt. Lett.* **24** 1805–7

[20] Lochab P, Senthilkumaran P and Khare K 2019 Propagation of converging polarization singular beams through atmospheric turbulence *Appl. Opt.* **58** 6335–45

[21] Dickson L D 1970 Characteristics of a propagating Gaussian beam *Appl. Opt.* **9** 1854–61

[22] Andrews L C, Phillips R L and Weeks A R 1997 Propagation of a Gaussian-beam wave through a random phase screen *Waves Random Media* **7** 229–44

[23] Ricklin J C, Miller W B and Andrews L C 1995 Effective beam parameters and the turbulent beam waist for convergent Gaussian beams *Appl. Opt.* **34** 7059–65

[24] Dios F, Rubio J A, Rodriguez A and Comeron A 2004 Scintillation and beam-wander analysis in an optical ground station-satellite uplink *Appl. Opt.* **43** 3866–73

[25] Recolons J, Andrews L C and Phillips R L 2007 Analysis of beam wander effects for a horizontal-path propagating Gaussian-beam wave: focused beam case *Opt. Eng.* **46** 18105–13

[26] Dowling J A and Livingston P M 1973 Behavior of focused beams in atmospheric turbulence: Measurements and comments on the theory *J. Opt. Soc. Am.* **63** 846–58

[27] Birch P, Ituen I, Young R and Chatwin C 2015 Long-distance Bessel beam propagation through Kolmogorov turbulence *J. Opt. Soc. Am.* A **32** 2066–73

[28] Bouchal Z 2002 Resistance of nondiffracting vortex beam against amplitude and phase perturbations *Opt. Commun.* **210** 155–64

[29] Gbur G and Tyson R K 2008 Vortex beam propagation through atmospheric turbulence and topological charge conservation *J. Opt. Soc. Am.* A **25** 225–30

[30] Aksenov V P and Kolosov V V 2015 Scintillations of optical vortex in randomly inhomogeneous medium *Photon. Res.* **3** 44–7

[31] Eyyuboğlu H T 2016 Scintillation behaviour of vortex beams in strong turbulence region *J. Mod. Opt.* **63** 2374–81

[32] Liu X and Pu J 2011 Investigation on the scintillation reduction of elliptical vortex beams propagating in atmospheric turbulence *Opt. Express* **19** 26444–50

[33] Paterson C 2005 Atmospheric turbulence and orbital angular momentum of single photons for optical communication *Phys. Rev. Lett.* **94** 153901

[34] Goyal S K, Ibrahim A H, Roux F S, Konrad T and Forbes A 2016 The effect of turbulence on entanglement-based free-space quantum key distribution with photonic orbital angular momentum *J. Opt.* **18** 064002

[35] Karimi E, Marrucci L, de Lisio C and Santamato E 2012 Time-division multiplexing of the orbital angular momentum of light *Opt. Lett.* **37** 127–9

[36] Leonhard N, Sorelli G, Shatokhin V N, Reinlein C and Andreas B 2018 Protecting the entanglement of twisted photons by adaptive optics *Phys. Rev.* A **97** 012321

[37] Schulz T J 2005 Optimal beams for propagation through random media *Opt. Lett.* **30** 1093–5

[38] Voelz D G and Xiao X 2009 Metric for optimizing spatially partially coherent beams for propagation through turbulence *Opt. Eng.* **48** 036001

[39] Lochab P, Senthilkumaran P and Khare K 2016 Near-core structure of a propagating optical vortex *J. Opt. Soc. Am.* A **33** 2485–90

[40] Tao R, Wang X, Si L, Zhou P and Liu Z 2013 Propagation of focused vector laser beams in turbulent atmosphere *Opt. Laser Technol.* **54** 62–7

[41] Andrews L C and Phillips R L 2005 *Laser Beam Propagation Through Random Media* (Bellingham, WA: SPIE Optical Engineering Press)

[42] Kogelnik H 1965 Imaging of optical modes—resonators with internal lenses *Bell Syst. Tech. J.* **44** 455–94

[43] Beckley A M, Brown T G and Alonso M A 2010 Full Poincaré beams *Opt. Express* **18** 10777–85

[44] Gibson C J, Bevington P, Oppo G-L and Yao A M 2018 Control of polarization rotation in nonlinear propagation of fully structured light *Phys. Rev.* A **97** 033832

[45] Churnside J H and Lataitis R J 1990 Wander of an optical beam in the turbulent atmosphere *Appl. Opt.* **29** 926–30

[46] Banakh V A, Krekov G M, Mironov V L and Khmelevtsov S S 1974 Focused-laser-beam scintillations in the turbulent atmosphere *J. Opt. Soc. Am.* **64** 516–8

[47] Titterton P J 1973 Scintillation and transmitter-aperture averaging over vertical paths *J. Opt. Soc. Am.* **63** 439–44

[48] Kerr J R and Eiss R 1972 Transmitter-size and focus effects on scintillations *J. Opt. Soc. Am.* **62** 682–4

Orbital Angular Momentum States of Light
Propagation through atmospheric turbulence
**Kedar Khare, Priyanka Lochab and Paramasivam Senthilkumaran**

# Appendix A

# Annotated computer code for beam propagation through turbulence

In this appendix we provide annotated code with brief additional explanations for simulation of arbitrary beam profiles through atmospheric turbulence by the split-step method discussed in chapter 8. The codes used here have been employed for generating a number of beam profile diagrams in this book. For beginning researchers as well as experts, the codes provided here can serve as a handy tool for experimentation and for trying out new ideas after suitable modification. The codes provided here work in the MATLAB programming environment as well as its Open Source clone GNU Octave. The high level programming languages used by both MATLAB and Octave are straightforward to follow and the reader is expected to know the basics of these programming tools that are now often used for teaching.

## A.1 Initialization of parameters

We first need to initialize all the variables for turbulence parameters, the two-dimensional sampling grid, wavelength, beam waist to be used, etc. These are defined below and are self-explanatory. The units used for each numerical quantity are specified in the comments that are shown after the % sign in each of the code lines.

### A.1.1 Turbulence parameters (scales and strength)

```matlab
Lo = 3;              % Outer turbulence scale [m]
lo = 0.01;           % Inner turbulence scale [m]
Cn2 = 1*10^(-14);    % Cn2value [m^(-2/3)]
```

## A.1.2 Sampling grid parameters

```
delx = 0.001;       % 2D sampling interval for computational
        grid [m].
% The sampling interval must be less than half of the
        inner scale (lo/2).

N = 512;       % Number of sample points along x and y
        directions

D = N*delx;       % Physical length of computational window [m].

% definition of (x,y) and spatial frequency (fx,fy)
   grids based on the
% sampling interval.

[x,y] = meshgrid(-D/2+delx/2:delx:D/2-delx/2);
[fx,fy] = meshgrid( -1/2/delx :   1/N/delx :   1/2/delx-1/N/delx );

% Polar coordinates
r = sqrt(x.^2+y.^2);
theta = atan2(y,x);
rho = sqrt(fx.^2 + fy.^2);
thetaf = atan2(fy,fx);
```

## A.1.3 Optical beam parameters

The parameters like wavelength, beam waist and propagation distance are defined next.

```
lambda = 1.5e-6;        % Wavelength [m]
k = 2*pi/lambda;        % Wave number [m^(-1)]
Wg = 0.1;               % Gaussian beam waist [m]
zo = (pi*Wg^2)/lambda;  % Rayleigh range
L = 2e3;                % Total propagation distance [m]
```

We need to estimate the total beam spread on propagation through turbulence of given strength over distance L. It must be ensured that the beam stays well within the computational window, otherwise the sampling grid parameters must change or the distance over which the propagation can be performed needs to be reduced. This is done via beam spread estimated using relations based on the Fried parameter.

```
rfried=(0.42*k∧2*Cn2*L)∧(-3/5);
n = (N*k*rfried∧2)/(4L);
```

The above relation for Fried parameter is a special case of equation (8.20) for constant $C_n^2$ over the propagation path. This relation may be modified suitably if required to accommodate variable turbulence strength. As per equation (8.23), we need to make sure that the quantity ($n$) above is greater than 4. As a second check, one may estimate the beam width after propagation by distance L through turbulence by finding the beam waist as per equation (8.29):

```
W = sqrt( Wg∧2 * (1 + (L/zo)∧2) + (2*L/k/rfried)∧2);
```

We can place a check here if the estimated beam spread 2W is well within the physical computational window size equal to N*delx. Generally if the beam has an initial concave curvature the beam will reduce in size till the focus distance and beam spreading beyond the computational window should not be an issue till the focusing distance is reached. If the above conditions are not valid, the sampling grid parameters or the distance L need modification. A Gaussian beam to be propagated through turbulence may be specified as:

```
u = exp(- r.∧2 / W∧2 − i*pi*r.∧2 / (lambda*F));
```

Here an additional concave curvature with radius of curvature F has been added on the beam in order to focus it. It has already been noted in equation (9.12) that in free space, a concave curvature with radius of $F$ leads to focusing of beam closer to the input plane. A focused vortex beam in the OAM state $l$ may be defined as:

```
uoam = A*r.∧l .* exp(- r.∧2 / W∧2 − i*pi*r.∧2 / (lambda*F)
+ i*l*theta);
```

The fields u (or uoam above may be normalized simply by using the Frobenium norm function, so that, they now have unit energy:

```
u = u / norm(abs(u), 'fro');
```

Other forms of initial fields (e.g. Bessel beams, flat-top beams, etc) may be defined in a straightforward manner by representing them over the (x,y) grid defined above. The field u will in general be a complex valued matrix of size N × N.

### A.1.4 Number of phase screens

The number of phase screens that are required for simulation is the next important part of the code. As discussed in section 8.1.3, point 5, we note that we need to place the screens periodically over a distance delz, such that, for this this distance the Rytov variance has a numerical value of less than 0.1 and also that the Rytov variance over the distance delz is less than 10% of the total Rytov variance over distance L. This can be achieved by slowly reducing the propagation distance L till both the conditions are satisfied. In the code below, the distance L is reduced in steps of deltaL = 10 m. Finally the number of screens are estimated as rounded up integer value of (L/L1) where L1 is the distance which satisfies both the required conditions.

```
totalrytov = 1.23*Cn2*k^(7/6)*L^(11/6);
flag1 = 0;
flag2 = 0;
L1 = L;
deltaL = 10;

while(flag1 == 0 || flag2 == 0)

L1 = L - deltaL;
rytovL1 = 1.23*Cn2*k^(7/6)*L1^(11/6);

if(rytovL1 < 0.1)
flag1 = 1;
end

if(rytovL1 < 0.1*totalrytov)
flag2 = 1;
end

end % End of while loop

numscreens = ceil(L/L1);
delz = L/numsreens;   % Distance between two random phase
        screens.
```

The distance L1 has been chosen as 10 m in the above code. It may be changed to a larger value in order to speed up this step.


## A.1.5 Generation of phase screen using FFT method

Having decided on the sampling grid and the number of phase screens needed based on the turbulence strength, we now need to generate the random phase screens that are required for the split-step method. The FFT method for phase screen generation has been explained in section 8.2.2. This method requires us to specify the phase spectrum and in the code provided below we use two choices, namely, Kolmogorov and von Karman. Their definition is as follows:

```
fo = 1/Lo;
fm = 5.92/(2*pi*lo);
A0 = 2*pi*k^2*delz;
% Kolmogorov Spectrum
spectrum1 = 0.033*Cn2*(rho).^(-11/3)*(2*pi)^(-11/3);
% von Karman Spectrum
spectrum2 = ...
0.033*Cn2*(rho.^2 + fo^2).^(-11/6).*(2*pi)^(-11/3).*exp(-rho/fm).^2);
```

The power spectral density for the FFT method is now defined as:

```
PSD = A0*spectrum;      % Here spectrum1 or spectrum2
                        % may be used.
PSD(N/2+1, N/2+1) = 0;  % Turning dc component to zero.
```

Now we generate the actual phase screen as follows:

```
Phi = 2*pi/(N*delx) * (randn(N,N) + 1i*randn(N,N)) .*
(PSD).^(0.5);
phi = real(fftshift(ifft2(ifftshift(Phi))));
phasescreen = exp(i*phi);
```

The phase function generated with FFT method still needs the sub-harmonic component which is generated as explained next.

### A.1.6 Generation of sub-harmonic phase screen

We will generate phase screens up to a level of p as explained in section 8.2.3. The main idea is to upsample the zero frequency pixel and evaluate the sub-harmonic phase screen phisub as a local discrete Fourier transform over the upsampled frequency locations. The self-explanatory code is very similar to the code for FFT based screen generation.

```
phisub = zeros(N,N);
delf = 1/(N*delx);

for p = 1:3
delfnew = delf/(3)^p;
[fxnew, fynew] = meshgrid(-1:1);
fxnew = fxnew*delfnew;
fynew = fynew*delfnew;
rhonew = sqrt(fxnew.^2 + fynew.^2);
A0 = 2*pi*k^2 *delz;
spectrum2 = ...
0.033*Cn2*(rhonew.^2+fo^2).^(-11/6)*(2*pi)^(-11/3).*exp
```

```
(-(rhonew/fm).^2);
PSDsub = A0*spectrum2;
PSDsub(2,2) = 0;
Phisub = 2*pi*delfnew*(randn(3,3) + 1i*randn(3,3)).*sqrt
(PSDsub);
phisub1 = zeros(N,N);

for j = 1:9
phisub1 = phisub1 + Phisub(j)*exp(i*2*pi*(fxnew(j)*x +
fynew(j)*y));
end

phisub = phisub + phisub1;

end

phisub = real(phisub) - mean(real(phisub(:)));
```

The phase screen generated using the FFT based method and the sub-harmonic part are added to get the total phase screen at any given location along the propagation path. So we have:

```
phasescreen = exp(i*phi + i*phisub);
```

In cases when the propagation path is slanted or vertical, the numerical value of $C_n^2$ as the beam propagates will be variable and this aspect may be readily incorporated in the computation by using the appropriate $C_n^2$ in the definition of both the FFT based and sub-harmonic phase screens.

### A.1.7 Free space propagation between two random phase screens

The free space propagation of scalar field between two random phase screens may be evaluated using the angular spectrum method as explained below. We will denote by uinit the field arriving at a phase screen and by u the field after transmission through the phase screen.

```
u = uinit.*phasescreen;
U = fftshift(fft2(ifftshift(u)));
alpha = sqrt(k^2-4*pi^2*(fx.^2 + fy.^2));
H = exp(i*delz*alpha);
fmax = 1/lambda * 1/sqrt(1 + (2*delz/N/p)^2);
LPfilter = (fx.^2 + fy.^2 <= fmax^2);
U = U.*H.*LPfilter;
udelz = fftshift(ifft2(ifftshift(U)));
```

The field udelz acts as uinit for the next segment in the split-step method.

### A.1.8 Simulation of propagation of vector beams through turbulence

A vector beam may be thought of as a coherent superposition of two scalar beam functions in orthogonal polarizations. For example, two different scalar beams u1 and u2 may be propagated over long paths through turbulence by the split-step method defined above. Since the cross-talk between the polarizations is nominal in long range propagation through turbulence, the total irradiance profile of a vector beam is approximated by summing the irradiances for the two scalar parts.

```
Itotal = abs(u1final).^2 + abs(u2final).^2;
```

Here u1final and u2final are field profiles for the two scalar components obtained after propagation simulation using the split-step method. If the two scalar beams are derived from the same initial laser source, the state of polarization of the vector beam may be of interest. The state of polarization may be determined pixel-by-pixel using the computed numerical values of u1final and u2final. Finally the polarization basis used may be the $x$–$y$ (Cartesian basis) or the RCP–LCP basis (or any other orthogonal basis states) and the conversion between the two basis states may be performed by appropriate linear combinations of pixels of u1final and u2final. Note that the total irradiance of the vector beam as computed above is independent of the choice of polarization basis.

www.ingramcontent.com/pod-product-compliance
Ingram Content Group UK Ltd.
Pitfield, Milton Keynes, MK11 3LW, UK
UKHW051941150726
7214IPUK00020B/369